WASTEWATER TREATMENT CONCEPTS AND PRACTICES

WASTEWATER TREATMENT CONCEPTS AND PRACTICES

FRANCIS J. HOPCROFT

MP MOMENTUM PRESS

MOMENTUM PRESS, LLC, NEW YORK

Wastewater Treatment Concepts and Practices
Copyright © Momentum Press®, LLC, 2015.

First published by Momentum Press®, LLC
222 East 46th Street, New York, NY 10017
www.momentumpress.net

ISBN-13: 978-1-60650-486-4 (print)
ISBN-13: 978-1-60650-487-1 (e-book)

Momentum Press Environmental Engineering Collection

DOI: 10.5643/9781606504871

Cover and interior design by Exeter Premedia Services Private Ltd., Chennai, India

10 9 8 7 6 5 4 3 2 1

Printed in the United States of America

ABSTRACT

This book provides a concise presentation of the fundamental elements of wastewater treatment process design. It shows the reader where various authors and authorities differ in their interpretation of the fundamentals and offers multiple tables of data from which to select appropriate design parameters. This book is intended to be a process design reference book, not a detailed design manual or a textbook suitable for classroom use.

KEY WORDS

environmental engineering, process design, reference manual, wastewater treatment

CONTENTS

LIST OF FIGURES   xi

LIST OF TABLES   xiii

FOREWORD   xv

ACKNOWLEDGMENTS   xvii

PREFACE   xix

1	CHEMISTRY CONSIDERATIONS	1
1.1	Introduction	1
1.2	Elements, Compounds, and Radicals	1
1.3	Reactive Characteristics of Atoms	3
1.4	Molecules	6
1.5	Moles and Normality	6
1.6	Properties of Radicals	7
1.7	Ions	8
1.8	Inorganic Chemicals	8
1.9	Units of Measure	8
1.10	Milliequivalents	12
1.11	Reaction Rates or "Reaction Kinetics"	15
1.12	Reactions Common to Wastewater Treatment	20
1.13	Coagulation and Flocculation	24
1.14	Hardness of Water	26
1.15	Chemical Oxygen Demand	26
1.16	Total Organic Carbon	27
1.17	Fats, Oil, and Grease	28

1.18	Material Balance Calculations	29
1.19	Emerging Chemicals of Concern	31
2	**BIOLOGY CONSIDERATIONS**	**35**
2.1	Introduction	35
2.2	Bacteria	35
2.3	Viruses	37
2.4	Algae	37
2.5	Fungi	40
2.6	Protozoans	40
2.7	Microscopic Multicellular Organisms	40
2.8	Pathogens	41
2.9	Indicator Organisms	41
2.10	Biological Oxygen Demand	42
2.11	BOD Formulas of Concern	43
2.12	Biological Decay Rate—k	43
2.13	Nitrogenous BOD	48
2.14	Temperature Effects on k-rate	49
2.15	Biological Growth Curve Kinetics	50
2.16	Dissolved Oxygen Concepts, Measurement, and Relevance	51
2.17	Biological Nitrification and Denitrification	52
3	**WASTEWATER TREATMENT PROCESSES**	**55**
3.1	Introduction	55
3.2	Basic Design Parameters	56
3.3	Preliminary Treatment Units	61
3.4	Primary Treatment Units	73
3.5	Secondary Treatment	79
3.6	Sludge Management	120
3.7	Tertiary Treatment Units	129
3.8	Details of Disinfection of Wastewater	131
4	**SEDIMENTATION FUNDAMENTALS**	**137**
4.1	Introduction	137
4.2	Discrete Particle Sedimentation	137

4.3	Flocculant Particle Sedimentation	138
4.4	Weirs	138
4.5	Clarifier Design	139
4.6	Flocculator Clarifiers	143
4.7	Design of Discrete Particle Clarifiers	143
4.8	Design of Flocculant Particle Clarifiers	150
4.9	Hindered Settling	153
4.10	Design of Upflow Clarifiers	154
5	**SUBSURFACE WASTEWATER DISPOSAL**	**157**
5.1	Introduction	157
5.2	Conventional Subsurface Disposal Systems	157
5.3	Alternative Disposal Field Designs	167
6	**WATER REUSE**	**175**
6.1	Introduction	175
6.2	Specific Reuse Options	175
	ABOUT THE AUTHOR	**183**
	INDEX	**185**

LIST OF FIGURES

Figure 1.1.   (a) Concentration versus time for Zero-Order
               Reactions and (b) Reaction Rate versus
               Concentration for Zero-Order Reactions.   16

Figure 1.2.   (a) Concentration versus time for First-Order
               Reactions and (b) Reaction Rate versus
               Concentration for First-Order Reactions.   17

Figure 1.3.   (a) Concentration versus time for Second-Order
               Reactions and (b) Reaction Rate versus
               Concentration for Second-Order Reactions.   18

Figure 2.1.   Typical BOD curve.   45

Figure 2.2. $(Time/BOD)^{1/3}$.   46

Figure 2.3.   Carbonaceous BOD with nitrogenous BOD curve.   49

Figure 2.4.   Typical growth curves for bacteria.   50

Figure 3.1.   Cross-sectional view of a typical Parshall flume. 68

Figure 3.2.   Schematic Views of a Rectangular (a) and
               a Circular (b) Clarifier.   77

Figure 3.3.   Slime build-up on a fixed film media.   80

Figure 3.4.   A classic trickling filter using a stone media bed
               (Adapted from: www.citywatertown.org/ and other
               places).   81

Figure 3.5.   First stage or single stage BOD_5 loadings plotted
               against removal efficiency for various recirculation
               factors. Based on Droste equations.   92

Figure 3.6.   Second stage BOD_5 loadings plotted against
               removal efficiencies for various recirculation factors.
               Based on Droste equations and assumes that BOD_5
               loadings have been corrected using Equation 3.18. 93

Figure 3.7.   Tapered aeration (top) versus step feed (bottom) –
               schematic only.   114

Figure 4.1.   Classic representation of the influent zone,
               settling zone, and effluent zone of a rectangular
               settling basin.   145

Figure 4.2.   Representative plot of the terminal velocities of
               various sized particles in a wastewater particles.   147

Figure 4.3.   Hypothetical settling efficiency plotted as a
               percentage of particles removed by depth as a
               function of time.   151

Figure 4.4.   Schematic of a typical upflow clarifier.   154

Figure 5.1.   Schematic of a typical septic tank and leaching
               field arrangement.   158

Figure 5.2.   Standard septic tank (not to scale).   158

Figure 5.3.   Distribution box (D-Box) (typ.) (not to scale).   159

Figure 5.4.   Typical leaching trench (not to scale).   159

LIST OF TABLES

Table 1.1.   Common elements and their common valence
             values   5

Table 1.2.   Common radicals (or, more accurately, polyatomic
             ions) and their electrical charge   7

Table 1.3.   Common elements, chemicals, radicals, and
             compounds with symbol or chemical formula,
             molecular weight, and equivalent weight   9

Table 1.4.   Emerging chemicals of concern   32

Table 2.1.   Pathogens often excreted by, or ingested by,
             humans   38

Table 2.2.   Maximum DO concentration with temperature   51

Table 3.1.   Comparative composition of raw wastewater   58

Table 3.2.   EPA discharge limits for wastewater treatment
             plants (40 CFR 133.102)   60

Table 3.3.   EPA discharge limits for wastewater treatment
             plants eligible for treatment equivalent to secondary
             treatment (40 CFR 133.105)   61

Table 3.4.   Typical design criteria for various preliminary
             wastewater treatment plant unit operations   63

Table 3.5.   Typical characteristics of domestic septage   72

Table 3.6.   Typical design parameters for primary clarifiers   78

Table 3.7.   Common characteristics of untreated, settled,
             primary sludge   79

Table 3.8.   Typical secondary system design parameters   84

Table 3.9.   Comparison of NRC equation variations for
             trickling filter design   87

Table 3.10.   First and second stage trickling filter efficiencies
               for BOD_5 loadings in lbs/1,000 cf/day or kg/m³/day
               for reactors of 1,000 cf or 1 m³ and the recirculation
               factors shown   89

Table 3.11.   Trickling filter design and performance parameters   99

Table 3.12.   Typical loading and operational parameters for
               secondary treatment processes   108

Table 3.13.   Design parameters for treatment ponds and
               lagoons   119

Table 3.14.   Design and performance parameters for gravity
               thickeners   122

Table 3.15.   Typical design and performance parameters for
               alternative sludge thickening options   124

Table 4.1.   Typical design characteristics of sedimentation
               basins and clarifiers   140

Table 5.1.   Wastewater design flows from various sources for
               subsurface disposal systems   160

Table 5.2.   Recommended soil loading rates in gpd/sf (L/m²/d)
               for various soil types   166

FOREWORD

This is the second in a series of volumes designed to assist senior level college students and graduate students with mastering the principles of environmental engineering. The premise behind these volumes is that it should not be necessary to peruse multiple volumes, technical papers, and textbooks to find the principles needed to comprehend various environmental engineering concepts. The intent is to include within one volume all the key principles needed to fully understand the concepts in a specific area of environmental engineering. It is assumed that the reader has at least a rudimentary understanding of basic chemistry, hydraulics, and fluid mechanics.

This volume addresses the narrow area of wastewater treatment. Other volumes in this series address water treatment, air pollution control, environmental chemistry, hydraulics, stormwater and Combined Sewer Overflow (CSO) control, lagoons, ponds, manmade wetlands, and other areas of environmental engineering. Those volumes are stand-alone references that address the key principles involved in each specific area. It may be necessary to refer to more than one volume to find a suitable solution to a complex problem, but if the student can dissect the problem and parse it into its fundamental components, it should be possible to find a specific volume in this series that will address the key principles of each component part.

It is the intent of this volume to address the process of wastewater treatment, not the mechanics of the machinery and reactors used to do the work. No amount of machinery and reactor vessels will ever treat wastewater effectively unless the process of using the equipment is properly developed first and is properly utilized afterwards.

ACKNOWLEDGMENTS

The editorial assistance of the following people is gratefully acknowledged:

Armen Casparian—Professor, Wentworth Institute of Technology (Retired)
Gautham Das—Professor, Wentworth Institute of Technology
Charles Pike—Black and Veatch Engineers, Boston, MA
Gergely Sirokman—Professor, Wentworth Institute of Technology
Paloma Valverde—Professor, Wentworth Institute of Technology
Grant Weaver—The Water Planet Company, New London, CT

PREFACE

The fundamental objective of wastewater treatment is to reduce the concentration of contaminants in the wastewater to a sufficiently low value that safe discharge to a receiving water, either a surface water or a sub-surface groundwater, can be accomplished. Achieving that goal requires the application of several fundamental principles of engineering. Among those are chemistry, biology, hydraulics, fluid mechanics, and mathematics of varying types. This book provides a synopsis of the basic fundamentals of those disciplines and then an outline of the use of those principles to solve specific wastewater engineering problems. This is intended as a process design and unit operation design reference manual, however, not a physical plant design reference manual.

Along with the various technical fields outlined in the previous paragraph, the effective discharge of properly treated wastewater also depends upon compliance with various federal, state, and local regulations. The selection of specific regulations to be consulted often revolves around the discharge location, rather than the actual treatment process. Nevertheless, it is important to consider all such regulatory frameworks when designing a wastewater treatment facility for any discharge. Several federal regulations of importance are discussed in this text as they arise in the discussion. Most notable are the secondary wastewater discharge permit limitations imposed by federal regulation, but implemented and enforced by the states, in most cases.

In addition, federal and state regulators have become increasingly concerned about issues of emerging contaminants such as personal care products and medicines that are persistent in the environment, resistant to treatment in the treatment plant, and harmful to humans and animals when discharged to the environment. Certain nutrients, such as nitrogen and phosphorous have long concerned regulators and need to be addressed proactively with all treatment processes. Various microbial contaminants have emerged as potentially significant health risk factors, and new indicator organisms that can be tracked easily have been identified for inclusion in future discharge permits.

As a result of all of these evolving concerns, the treatment of wastewater is an evolving science and a lot of new ways to accomplish the old objectives are constantly under development. A separate volume will address new and emerging technologies, and it is expected that this volume will be updated regularly to cover those changes to the practice of wastewater treatment.

CHAPTER 1

CHEMISTRY CONSIDERATIONS

1.1 INTRODUCTION

A fundamental understanding of chemistry is an important part of understanding how wastewater treatment works. This is not a subject commonly favored by civil and environmental engineers. Fortunately, it is not necessary to be a chemist in order to be effective at designing suitable wastewater treatment processes, although a basic understanding of biochemistry and microbiology is very helpful.

1.2 ELEMENTS, COMPOUNDS, AND RADICALS

In the first section, a review of the key points of chemistry necessary for effective wastewater treatment is presented and discussed. This review includes discussions of elements, ions, radicals, and compounds. It is assumed that the reader is reasonably familiar with the notion that atoms are made up of electrons, protons, and neutrons important to physics, but perhaps less important to wastewater treatment. These atoms are the basic building blocks of all the other forms of chemical structures.

Elements are made up of atoms. The number of atoms in a group determines how much of the element is present, but even one atom, properly constructed by nature, constitutes an element. There is a limited number of ways that electrons, protons, and neutrons can combine to form atoms. Whenever they do combine into a stable form, a different element is created. There are only a few elements that are used in wastewater treatment in their pure form. Chlorine gas and ozone gas are two examples. Oxygen is required as a separate element, but is seldom applied in pure form. It is noted that finding any element in a truly pure form is hard to do except in a high purity laboratory. Most wastewater treatment needs do not require absolute purity, and essentially all elements are provided and applied in some form of compound with other elements at various degrees of purity. It is essential to verify the purity of the elements within the compounds before calculating masses of material to use in the field.

Compounds are made up of various elements joined together by electrical charges into stable structures; although, some are known to be much more stable than others. They are considered to be electrically neutral and all component atoms have their desired number of electrons with none left over for further interactions. Most of the chemicals used in wastewater treatment are made up of various compounds. Ferric chloride, used to assist with precipitation; sodium hydroxide, used for pH control; and potassium permanganate, used for odor control, are examples of some of the many compounds used in wastewater treatment. Each is made up of two or more elements chemically combined into a stable compound. It is most common to find that the compounds used are no more pure than the elements. For example, the actual concentration of permanganate in a sample of potassium permanganate can, in practice, vary. In principle, the ratio of the potassium ions to permanganate ions is fixed, but impurities can contaminate the sample depending on the reagent grade selected. Nevertheless, it is important, when calculating quantities of compounds to use, to ensure that the actual concentration of the desired compound in the mix is known or determined.

Ions are charged atoms or groups of atoms. If they are positively charged, because they have fewer than expected electrons, they are called cations. Metal atoms, such as calcium or iron, tend to lose one or more electrons and commonly form cations. If they are negatively charged, because they have gained one of more electrons, they are called anions. Nonmetals, such as oxygen and chlorine, tend to gain one or more electrons and commonly form anions. Oxoanions, such as chlorites and sulfates, are quite common, due to the ubiquitous and reactive nature of oxygen. They are chemical combinations (chemical bonds) of oxygen and another nonmetal but behave as a single anion. Ions form because they have more stable electron configurations than the neutral atoms, especially when metal elements find themselves in physical contact with nonmetal elements.

Radicals are chemical species that have unpaired electrons. They are sometimes electrically neutral, sometimes not. (There is often some confusion in the nomenclature between ions and radicals.) Radicals are unstable due to having unpaired electrons. Therefore, radicals are much more likely to react with other chemical elements, radicals, or compounds. Radicals are in fact always looking for something to react with so that they can become electrically neutral. This is a rather convenient characteristic of radicals when reactions with them are desired, but a very difficult characteristic to control when other reactions are desired preferentially to those involving the specific radical in question. (See Section 1.6 for more on radicals.)

1.3   REACTIVE CHARACTERISTICS OF ATOMS

1.3.1   ATOMIC WEIGHT

Each atom is made up of a unique combination of electrons, protons, and neutrons. Changing the number of protons will change the characteristics of the atom and convert it to a different element. In addition, each of those electrons and protons has a mass. The mass of an electron is certainly very small. The majority of mass arises from protons and neutrons, and with the range in variation of those particles, some atoms can become very heavy relative to other atoms. In fact, each atom has a specific mass called the "atomic weight" of that atom. Since it is very difficult for most wastewater designers to actually weigh an atom of anything, it is customary to define the weight of each atom relative to the weight of hydrogen. That is because hydrogen contains exactly one electron and one proton. Consequently, every other element contains more than one of each and therefore must weigh more than hydrogen. An atom of helium, which contains two electrons, two protons, and two neutrons (neutrons have approximately the same mass as a proton but they have no charge; they do not change the chemical properties of an atom, but they are responsible for creating different, naturally occurring isotopes of a given element), for a total mass of four, must weigh exactly four times as much as a hydrogen atom and therefore it is assigned an atomic weight of 4. Helium four (^{4}He) is the most common isotope of helium and hence has the assigned mass of 4.

An atom of helium contains two electrons, two protons, and two neutrons (electrons and protons always occur in pairs, since the electrons are electrically negative and the protons are electrically positive, yielding an electrically neutral atom). Protons and neutrons have nearly the same mass; therefore, a helium atom must weigh nearly four times as much as a hydrogen atom and is assigned an atomic weight of 4.

In 1961, however, the standard was actually changed such that the standard became the ^{12}C atom, or carbon 12, isotope. This isotope was defined to have an atomic weight of exactly 12, relative to hydrogen. This changes the atomic weights of all the other elements very slightly;

for example, hydrogen now sits at 1.008 (rounded to 1 for all practical purposes) and oxygen has an atomic weight of 15.9994 under the carbon 12 standard (rounded to 16.0 for all practical purposes). The atomic weight of carbon on a periodic table is generally shown as 12.011, even though that is the standard by which the other weights are measured. The reason for that is that there are several variations, or isotopes, of many of the atoms, including carbon. The atomic weight shown on the periodic table is the average of the atomic weights of all the isotopes. Thus, although the carbon 12 isotope does weigh exactly 12, the others do not and the average of all the isotopes is slightly higher.

The atomic weight of an atom, then, is actually its weight relative to the weight of a carbon 12 atom. Work done prior to 1961 in which the atomic weights were carefully measured or used may show slightly different experimental results than work done subsequent to the change in standard and that should be considered when comparing historical data to current data.

1.3.2   GRAM ATOMIC WEIGHT

It is noted, however, that even though it is possible to indicate the atomic weight of an element, that value is often difficult to use because it has no units. Atomic mass was apparently not originally defined in unitless terms. It was specifically defined in units of grams/mole. Atomic mass units may also be used at times. In any case, it has now become customary to define the atomic weight of an element in terms of "gram atomic weight" of that element and to define the ratio in grams. The gram atomic weight of hydrogen, then, is 1 and that of helium is 4. (The more precise atomic weight of hydrogen is 1.008 and that of helium is 4.003. For purposes of calculations, gram atomic weights are generally used as whole numbers.) The gram atomic weight of every other element is calculated in the same way.

1.3.3   VALENCE (ALSO KNOWN AS "OXIDATION STATE")

The concept of valence is a measure of the ability of an element to combine with other elements. Even stable elements will give up an electron or share an electron with another element under the right conditions. Valence indicates the combining power of an element relative, again, to that of hydrogen. Hydrogen has one electron, so it has a combining power of one. An element with two electrons has the potential to have a combining power of two, and so forth. However, not all electrons are always available for combining. Electrons typically occupy specific orbits around the protons and neutrons and only those in the outermost orbits are available for combining. That generally limits the combining power to no more than six, regardless of how many electrons an element may contain.

In general, a plus valence indicates that the element prefers to lose electrons and has the ability to replace hydrogen atoms in a compound when the two compounds react with each other, while a negative valence indicates that the element prefers to gain electrons and will react with hydrogen to form a new compound. Table 1.1 shows various elements common to wastewater treatment and their common valence values.

1.3.4 EQUIVALENT WEIGHT AND COMBINING WEIGHT

The concept of valence leads to one more weight unit associated with elements. That unit is called the "equivalent weight" or the "combining weight" of the element. Each element has a unique equivalent weight equal to its atomic weight divided by its valence. Since each element has a unique atomic weight, but the valence is limited to a small number of

Table 1.1. Common elements and their common valence values

Aluminum	3^+	Lead	$2^+, 4^+$
Arsenic	3^+	Magnesium	2^+
Barium	2^+	Manganese	$0, 2^+, 3^+, 4^+, 6^+, 7^+$
Boron	3^+	Mercury	$1^+, 2^+$
Bromine	1^-	Nickel	2^+
Cadmium	2^+	Nitrogen	$3^-, 0, 1^+, 2^+, 3^+, 4^+, 5^+$
Calcium	2^+	Oxygen	2^-
Carbon	$4^-, 3^-, 2^-, 1^-, 0, 1^+, 2^+, 3^+, 4^+$	Phosphorous	5^+
Chlorine	$1^-, 0, 1^+, 3^+, 4^+, 5^+, 7^+$	Potassium	1^+
Chromium	$3^+, 6^+$	Selenium	6^+
Copper	$1^+, 2^+$	Silicon	4^+
Fluorine	1^-	Silver	1^+
Hydrogen	1^+	Sodium	1^+
Iodine	1^-	Sulfur	$2^-, 0, 2^+, 4^+, 6^+$
Iron	$2^+, 3^+$	Tin	$2^+, 4^+$
		Zinc	2^+

integers, it is possible for more than one element to have the same equivalent weight and for the same element to have more than one equivalent weight. This should not present any difficulty in calculations.

1.4 MOLECULES

When atoms get together they form molecules. The term "element" can be used as a collective term for a type of atom. Similarly, the term "compound" can be used as a collective term for a type of molecule. The molecules of a compound are as unique as the atoms that form the elements that comprise the molecule. Not surprisingly, each molecule has a molecular weight (also known as the "molar mass") equal to the sum of the atomic weight of the elements that form the molecule. Thus, the molecular weight of water, comprised of two atoms of hydrogen and one atom of oxygen, is equal to two times the atomic weight of hydrogen ($1 \times 2 = 2$). Since there are two hydrogen atoms present, plus one times the atomic weight of oxygen ($1 \times 16 = 16$), because of the one oxygen atom present, the total molecular weight of water is, therefore, $2 + 16 = 18$.

1.5 MOLES AND NORMALITY

Due to the need for clarity when accounting for amounts of materials, a convention has evolved to use a different measure for molecules, called "moles." A mole is a unit of count, to allow for the tracking of very large numbers of individual particles. Much like a pair is 2, a dozen is 12, and a score is 20, a mole is 6.02×10^{23}. The number may be dauntingly large, but it functions the same way each of the other examples functions. One mole of anything is also equal to the sum of the gram atomic weights of the elements that make up that thing, expressed in grams. In essence, the gram molecular weight of a molecule is equal to one mole of that molecule. Therefore, one mole of oxygen is equal to 16 grams of oxygen and one mole of pure water is equal to 18 grams of pure water.

Example Problem 1.1 shows how the molecular and equivalent weights of compounds are related.

Example Problem 1.1

Calculate the molecular weight and the equivalent weight of calcium carbonate.

Solution

The formula for calcium carbonate is $CaCO_3$. Using the atomic weights from Table 1.1, the atomic weights of the elements are:

Ca	$1 \times 40.1 = 40.1$
C	$1 \times 12.0 = 12.0$
O	$3 \times 16.0 = 48.0$
Molecular Weight	$= 100.1$ or 100 grams

The calcium atom has a valence of 2^+ (see Table 1.1), while the carbonate ion has an electrical charge of 2^- (see Table 1.2). Thus the equivalent weight of the compound is the molecular weight (100) divided by the valence (2), or 50 grams per equivalent weight.

This concept then leads to the concept of Normality. A 1-Normal solution of a substance is equal to one equivalent weight, or one mole of charge, of that substance dissolved in sufficient water to create one liter of solution.

1.6 PROPERTIES OF RADICALS

As indicated earlier, radicals are groups of molecules that do not quite achieve the electrical stability of a compound, but which do usually contain atoms of more than one element. Radicals are not electrically stable because they have an odd number of electrons in their structure. This makes them much more prone to react with other molecules or radicals and much less stable than an electrically neutral compound or element. The lack of an electron in the radical generally means that the radical will have a positive charge, while an excess electron will generally yield a

Table 1.2. Common radicals (or, more accurately, polyatomic ions) and their electrical charge

Ammonium	1^+	Hypochlorite	1^-
Bicarbonate	1^-	Nitrate	1^-
Bisulfate	1^-	Nitrite	1^-
Bisulfite	2^-	Orthophosphate	3^-
Carbonate	2^-	Sulfate	2^-
Hydroxyl	1^-	Sulfite	2^-

negative charge. These electrical charges are similar to the valence of an atom discussed in Section 1.3.3. Table 1.2 shows various radicals important to wastewater treatment and their common electrical charge.

1.7 IONS

When inorganic compounds dissolve in water, and sometimes when they dissolve in other substances, they dissociate, or break down, and ionize into electrically charged atoms called "ions." Ions have an electrical charge that is also similar to that of valence or oxidation state, as discussed in Section 1.3.3. This means that these ions will also combine with other ions based on their combining power, or electrical charge. The objective of the ion is to become electrically neutral, so an ion with a charge of +3 will react easily with a different ion having an electrical charge of –3, or with three separate ions each having an electrical charge of –1.

1.8 INORGANIC CHEMICALS

Much of what has been discussed so far has to do with both organic and inorganic molecules and compounds. Organic compounds are defined (with a few exceptions) as those that contain carbon, while inorganic compounds are those that do not contain carbon. Organic and inorganic compounds tend to act differently under similar circumstances, so it becomes important to understand which type of compound is being discussed or used. Both types have similar properties of molecular weight and equivalent weight, as discussed in Sections 1.4 and 1.3.4. Table 1.3 provides a list of common chemicals with the symbol or chemical formula, molecular weight, and equivalent weight of the common form of each.

1.9 UNITS OF MEASURE

Atoms, ions, and compounds are generally reported in terms of a concentration in milligrams per liter (mg/L) of the element in a solute, usually water in wastewater treatment discussions. It is noted that this is a measure of concentration; it is not a measure of amount or level. The amount of a compound or chemical present is the total mass of that compound or chemical within the total volume of solute. Since the total volume of solute is a constantly changing variable in almost every wastewater treatment reactor, the amount of a material present at any given instant

Table 1.3. Common elements, chemicals, radicals, and compounds with symbol or chemical formula, molecular weight, and equivalent weight

Name	Symbol or formula	Atomic or molecular weight	Equivalent weight
Activated Carbon	C	12.0	N/A
Aluminum	Al	27.0	9.0
Aluminum Hydroxide	$Al(OH)_3$	78.0	26.0
Aluminum Sulfate	$Al_2(SO_4)_3 \cdot 14.3\ H_2O$	600	100
Ammonia	NH_3	17.0	N/A
Ammonium	NH_4^+	18.0	18.0
Ammonium Fluosilicate	$(NH_4)_2SiF_6$	178	N/A
Ammonium Sulfate	$(NH_4)_2SO_4$	132	66.1
Arsenic	As	74.9	25.0
Barium	Ba	137.3	68.7
Bicarbonate	HCO_3^-	61.0	61.0
Bisulfate	HSO_4^-	97.0	97.0
Bisulfite	HSO_3^-	81.0	81.0
Bromide	Br^-	79.9	79.9
Cadmium	Cd	112.4	56.2
Calcium	Ca	40.1	20.0
Calcium Bicarbonate	$Ca(HCO_3)_2$	162.0	81.0
Calcium Carbonate	$CaCO_3$	100.0	50.0
Calcium Chloride	$CaCl_2$	111.1	55.6
Calcium Fluoride	CaF_2	78.1	N/A
Calcium Hydroxide	$Ca(OH)_2$	74.1	37.0
Calcium Hypochlorite	$Ca(OCl)_2 \cdot 2H_2O$	179	N/A
Calcium Oxide	CaO	56.1	28.0
Carbon	C	12.0	N/A
Carbonate	CO_3^{2-}	60.0	30.0
Carbon Dioxide	CO_2	44.0	22.0
Carbon Monoxide	CO	28.0	14.0
Chlorine	Cl	35.5	35.5
Chlorine Dioxide	ClO_2	67.0	N/A

(Continued)

Table 1.3. (*Continued*)

Name	Symbol or formula	Atomic or molecular weight	Equivalent weight
Chromium	Cr	52.0	17.3
Common (Table) salt	NaCl	58.4	58.4
Copper	Cu	63.5	31.8
Copper Sulfate	$CuSO_4$	160	79.8
Copperas	$FeSO_4 \cdot 7H_2O$	278	139
Ferric Chloride	$FeCl_3$	162	54.1
Ferric Hydroxide	$Fe(OH)_3$	107	35.6
Ferric Sulfate	$Fe_2(SO_4)_3$	400	66.7
Ferrous Sulfate	$FeSO_4 \cdot 7H_2O$	278	139
Hydrochloric Acid	HCl	36.5	36.5
Hydrogen	H	1.0	1.0
Hydrogen Sulfide	H_2S	34.1	
Hydroxyl	OH^-	17.0	17.0
Hydrated Lime	$Ca(OH)_2$	74.1	37.0
Hypochlorite	ClO^-	51.5	51.5
Iron	Fe	55.8	27.9
Lead	Pb	207.2	103.6
Lime (Calcium Oxide)	CaO	56.1	28.0
Magnesium	Mg	24.3	12.2
Magnesium Hydroxide	$Mg(OH)_2$	58.3	29.2
Magnesium Sulfate	$MgSO_4$	120	60.2
Manganese	Mn	54.9	27.5
Mercury	Hg	200.6	100.3
Methane	CH_4	16.0	16.0
Methanol	CH_4O (or CH_3OH)	32.0	N/A
Nickel	Ni	58.7	29.4
Nitrate	NO_3^-	62.0	62.0
Nitrite	NO_2^-	46.0	46.0
Nitrogen	N	14.0	N/A
Orthophosphate	PO_4^{3-}	95.0	31.7

(*Continued*)

Table 1.3. (*Continued*)

Name	Symbol or formula	Atomic or molecular weight	Equivalent weight
Oxygen	O	16.0	16.0
Ozone	O_3	48.0	N/A
Potassium	K	39.1	39.1
Potassium Permanganate	$KMnO_4$	158	N/A
Selenium	Se	79.0	13.1
Silver	Ag	107.9	N/A
Soda Ash	$NaCO_3$	106	107.9
Sodium	Na	23.0	53.0
Sodium Bicarbonate	$NaHCO_3$	84.0	N/A
Sodium Bisulfite	$HNaO_3S$	104	N/A
Sodium Carbonate	$NaCO_3$	106	84.0
Sodium Chloride	NaCl	58.4	53.0
Sodium Fluoride	NaF	42.0	58.4
Sodium Fluorosilicate	Na_2SiF_6	188	N/A
Sodium Hydroxide	NaOH	40.0	N/A
Sodium Hypochlorite	NaOCl	74.4	40.0
Sodium Silicate	Na_4SiO_4	284.0	N/A
Sodium Thiosulfate	$Na_2S_2O_3$	158.0	N/A
Sulfite	SO_3^{2-}	80.0	40.0
Sulfate	SO_4^{2-}	96.0	48.0
Sulfur	S^{2+}	32.1	N/A
Sulfur Dioxide	SO_2	64.1	32.0
Sulfuric Acid	H_2SO_4	98.1	16.0
Zinc	Zn^{2+}	65.4	N/A

is generally not important or even relevant. Similarly, a level refers to a vertical distance from a horizontal reference point. The top of a sludge layer, or "blanket," may have a level to it if it accumulates in the bottom of a reactor and begins to fill the reactor. The sludge blanket level would then refer to the distance from the bottom of the reactor to the top of the sludge blanket. The amount, then, refers to the total mass of the compound present, the level refers to the location within a vertical plane, and the concentration refers to the amount or mass of the substance per unit of volume (typically 1 L). A gram, or milligram, is a unit of mass and a liter is a unit of volume, hence the term mg/L is a measure of concentration, not a measure of either amount or level.

1.10   MILLIEQUIVALENTS

Milliequivalents (meq) are used in two distinct ways. The first involves converting the data developed during titrations into usable units of weight. It happens, however, that the standard unit of measure during titration is a volume measure (milliliters, or mL), not units of mass. Therefore, the mass per milliliter of titrant must be known to convert the units properly. As indicated in Section 1.5, one mole of a substance dissolved in 1 L of water equals a "1-Normal" concentration of that material. Similarly, two moles of a substance in 1 L of water would yield a 2-Normal solution. Consequently, when normal solutions are being used, the equation for milliequivalents, in units of volume, is the following.

$$\text{mL of titrant} \times N = \text{meq of active material in the titrant} \qquad (1.1)$$

This also means that the meq of active material in the titrant used is equal to the meq of active material in the sample being titrated. To convert those data to a concentration of active material in the sample, it is necessary to know the volume of the sample being titrated.

$$\begin{aligned} &\text{meq/L of active material in the sample} = \\ &\qquad (\text{mL of titrant} \times N \times 1000)/(\text{sample volume in mL}) \qquad (1.2) \end{aligned}$$

It is more common, however, to report the concentration in terms of mass than in terms of volume. To convert the volumetric measure to a mass measure, the follow equation is used:

$$\begin{aligned} &\text{mg/L of active material in sample} = (\text{mL of titrant} \times N \\ &\qquad \times \text{Equivalent Weight} \times 1000)/(\text{sample volume in mL}) \qquad (1.3) \end{aligned}$$

Sometimes it is desirable to indicate the combining weight of a substance, similar to the combining weight of an element, as discussed in Section 1.3.4. In this case, the concept of milliequivalents per liter, or meq/L, is also used on a mass basis. Milliequivalents are calculated slightly differently depending upon whether they are being calculated for compounds or polyatomic ions. In the end, however, they are always equal to the concentration of the compound or radical divided by the equivalent weight of that compound or radical.

Equation 1.4 shows the calculation of milliequivalents for compounds and Equation 1.5 shows the calculation of milliequivalents for radicals.

The concentration of an ion in solution can be expressed in meq/L, which represents the combining weight of the ion, radical, or compound. The meq/L is calculated from the concentration in milligrams per liter (mg/L) by Equation 1.4.

$$\begin{aligned} \text{meq/L} &= (\text{mg/L}) \times (\text{valence/atomic weight}) \\ &= (\text{mg/L})/(\text{equivalent weight}) \end{aligned} \qquad (1.4)$$

In the case of a radical or compound, the equation is slightly different, as shown by Equation 1.5. The difference is in the mechanism for calculating the equivalent weight.

$$\begin{aligned} \text{meq/L} &= (\text{mg/L}) \times (\text{electrical charge/molecular weight}) \\ &= (\text{mg/L})/(\text{equivalent weight}) \end{aligned} \qquad (1.5)$$

Milliequivalents are used to check the chemistry of treated wastewater and to help decide how much of a particular chemical (in concentration units of mg/L) should be added to the treated water to yield specific desired results. Example Problem 1.2 shows how this is done.

Example Problem 1.2

Assume that an analysis of a water sample shows the following results.

Calcium	32.0 mg/L	Magnesium	15.8 mg/L
Sodium	23.0 mg/L	Potassium	13.9 mg/L
Bicarbonate	173.0 mg/L (as HCO_3)	Sulfate	35.0 mg/L
Chloride	24.5 mg/L		

Changing mg/L concentrations to meq/L concentrations, identify the hypothetical chemical combinations that should result in this water. If a different concentration of any resultant compound is desired, the concentration of each of the components listed needs to be adjusted to create the target concentration of the desired component.

Solution

Set up a table, using Equation 1.4, as follows.

Component	Valence	Concentration in mg/L	Equivalent weight	Concentration in meq/L
Ca	2+	32.0	20.0	1.60
Mg	2+	15.8	12.2	1.30
Na	1+	23.0	23.0	1.00
K	1+	13.9	39.1	0.36
			Total cations	4.26
HCO_3	1−	173.0	61.0	2.84
SO_4	2−	35.0	48.0	0.73
Cl	1−	24.5	35.5	0.69
			Total anions	4.26

The hypothetical combinations that could occur from these concentrations are shown in the following table.

Hypothetical combinations	Hypothetical concentrations in meq/L
$Ca(HCO_3)_2$	1.60
$Mg(HCO_3)_2$	1.24
$MgSO_4$	0.06
$Na_2(SO_4)$	0.67
NaCl	0.33
KCl	0.36

This is shown graphically in the following chart.

Ca^{+2} @ 1.60 meq/L		Mg^{+2} @ 1.30 meq/L	Na^{+1} @ 1.00 meq/L		K^{+1} @ 0.36 meq/L
HCO_3^{-1} @ 2.84 meq/L			SO_4^{-2} @ 0.73 meq/L	Cl^{-1} @ 0.69 meq/L	
$Ca(HCO_3)_2$ @ 1.60 meq/L		$Mg(HCO_3)_2$ @ 1.24 meq/L	$Na_2(SO_4)$ @ 0.67 meq/L	NaCl @ 0.33 meq/L	KCl @0.36 meq/L

$MgSO_4$ @ 0.06 meq/L

As noted earlier, most compounds and chemicals used in wastewater treatment are not pure. This means that the amount to be added to achieve a specific desired outcome must be adjusted to account for the impurities present. Example Problem 1.3 shows how that is done.

Example Problem 1.3

The equation for the removal of calcium hardness from water using the lime precipitation process is described as follows:

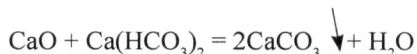

$$CaO + Ca(HCO_3)_2 = 2CaCO_3 \downarrow + H_2O$$

Given lime with a purity of 68 percent CaO, what dosage of lime is needed to precipitate 75 mg/L of calcium?

Solution

1 mole of $Ca(HCO_3)_2$ has a gram molecular weight of 162 grams, but contains 40.1 grams of calcium. 75 grams of calcium is equivalent to:

(75 mg/L of Ca^{2+}) (162 g/mole of $Ca(HCO_3)_2$/(40.1 grams of Ca^{2+} per mole of $Ca(HCO_3)_2$) = 383 mg/L $Ca(HCO_3)_2$

1 mole of CaO has a gram molecular weight of 56.1 grams. This mole will combine with one mole of $Ca(HCO_3)_2$, or 162 grams of $Ca(HCO_3)_2$.
383 mg/L of Ca $(HCO_3)_2$ will react with:

(56.1 grams CaO/162 grams $Ca(HCO_3)_2$) × 383 mg/L $Ca(HCO_3)_2$
= 132.6 mg/L CaO

For a purity of 68 percent, the dosage of the available lime required is:

(132.6 mg/L)/0.68 = 195 mg/L

1.11   REACTION RATES OR "REACTION KINETICS"

The rate or speed with which a chemical reaction occurs is not universally constant. Some reactions occur very quickly, while others occur very slowly. Still others take a moderate amount of time to occur. In addition, some reactions require an input of energy, usually in the form of heat, to work, while others give off heat, often copious quantities of heat and often very quickly, during their reactions.

The rate at which a reaction occurs is defined by various "orders" of reaction, which generally depend upon whether the reaction rate is driven by the mere presence of a compound or whether the concentration present is important. The order of the reaction depends on what factors control the rate and determine the resulting concentration of the reactants over time.

1.11.1 ZERO ORDER REACTIONS

A "zero order" reaction depends only on the presence of the reactant, not the concentration. Any amount of reactant present will cause the reaction to proceed. This type of reaction generally proceeds at a constant rate, once it starts, until the entire mass of reactant has been totally consumed by the reaction. This is shown graphically in Figure 1.1 (a). It is noted that the slope of the remaining concentration line over time on that graph is defined as "k," which is the reaction rate constant, or the constant rate at which this reaction occurs. The units of k are in 1/time, typically 1/days, or 1/d. The equation of this line is defined by Equation 1.6.

$$C = C_o - kt \qquad (1.6)$$

Where:

C = concentration of the reactant at any time, t
C_o = concentration of the reactant at time, t = 0
k = reaction rate constant in units of d^{-1}
t = time since the start of the reaction in days

Figure 1.1 (b) shows the reaction rate as a function of the concentration. It is noted that with a zero-order reaction, in which the reaction rate is unrelated to the concentration, that line is flat.

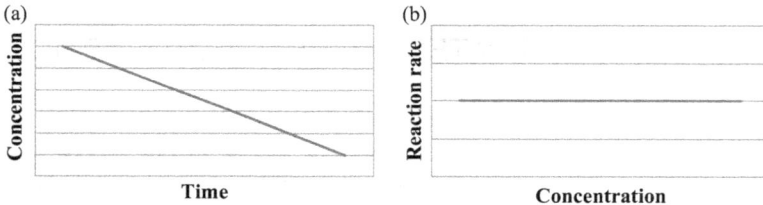

Figure 1.1. (a) Concentration versus time for Zero-Order Reactions and (b) Reaction Rate versus Concentration for Zero-Order Reactions.

1.11.2 FIRST ORDER REACTIONS

Most wastewater treatment reactions do depend upon the concentration of the reactant, however, which makes those reactions "first order" reactions. First order reactions yield a curved graph of concentration over time, as shown in Figure 1.2(a). The shape of the concentration plot over time on this graph is a curve because the concentration changes at a variable rate since the rate of the reaction slows as the concentration of the reactant decreases.

Rather than trying to deal with the equation for the slope of a curved line, however, it is most often easier to change the curved line to a straight one to make calculations easier. This is done by plotting the same data on semi-log graph paper, with concentration on a vertical log scale and time on a normal horizontal scale. Either the tangent to the curved line at any point, or the slope of the straight line at any point, will equal the value of k for the first order reaction. It is noted that this value is constantly changing, on both graphs, because the concentration is constantly changing and the reaction rate, k, is a function of the concentration. The equation for the first order reaction is shown as Equation 1.7.

$$C = C_o \, e^{-kt} \tag{1.7}$$

Where:

    C = concentration of the reactant at any time, t
    C_o = concentration of the reactant at time, t = 0
    e = mathematical e, approximately 2.71828
    k = reaction rate constant at the moment of measurement, in d^{-1}
    t = time since the start of the reaction in days

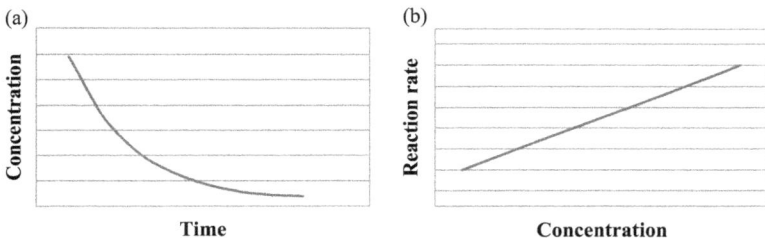

Figure 1.2. (a) Concentration versus time for First-Order Reactions and (b) Reaction Rate versus Concentration for First-Order Reactions.

Figure 1.2 (b) shows the reaction rate as a function of the concentration. It is noted that with a first-order reaction, in which the reaction rate is directly related to the concentration, that line is a straight line but that it also declines because the rate decreases linearly as the concentration decreases.

1.11.3 SECOND ORDER REACTIONS

Second order reactions occur at a rate dependent upon the square of the reactant concentration when that reactant is being converted to a single reaction product. Secondary reactions of the second order may also be occurring at different rates due to the presence of other reactants in the mix. This is shown graphically in Figure 1.3(a). It is noted that the slope of the remaining concentration line over time on that graph is defined as "k," which is the reaction rate constant, or the constant rate at which this reaction occurs. The equation for this type of reaction is the following.

$$1/C - 1/C_0 = kt \tag{1.8}$$

Where:

C = concentration of the reactant at any time, t
C_0 = concentration of the reactant at time, t = 0
k = reaction rate constant at the moment of measurement, in d^{-1}
t = time since the start of the reaction in days

Figure 1.3 (b) shows the reaction rate as a function of the concentration. It is noted that with a second-order reaction, in which the reaction rate is directly related to the square of the concentration, that line is an exponentially increasing curve on this graph because the concentration increases to the right.

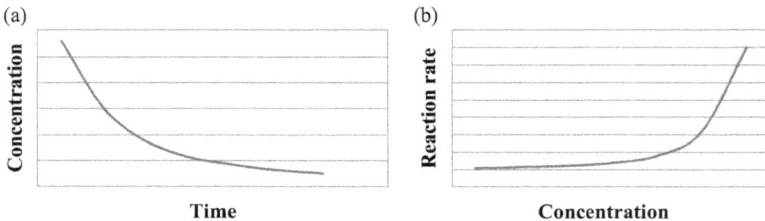

Figure 1.3. (a) Concentration versus time for Second-Order Reactions and (b) Reaction Rate versus Concentration for Second-Order Reactions.

1.11.4 THIRD AND FOURTH ORDER REACTIONS

Third and fourth order reactions also occur in nature, but they are extremely rare in wastewater treatment and are not included in this discussion.

1.11.5 EFFECTS OF TEMPERATURE ON THE VALUE OF k

In all cases, the reaction rate constant, k, is a function of temperature, which is why heating things generally causes reactions to proceed more quickly. This means that the value of k has to be adjusted if the temperature of the reactants is not within a "normal" value of approximately 20°C. A slight variation of a degree or two either side of that normal value will not yield a significant change in the value of k and is not likely to affect the way a treatment process proceeds. More than a one or two degree variation in the temperature, however, may affect the reaction and it should be checked.

The correction factor for the reaction rate constant is shown in Equation 1.9.

$$k_2 = k_1 \theta^{(t_2 - t_1)} \tag{1.9}$$

Where:

k_2 = the corrected reaction rate constant, d^{-1}
k_1 = the initially calculated reaction rate constant, d^{-1}
θ = a conversion rate constant, usually having the unitless value of 1.072
t_2 = the temperature at which the k factor is desired, °C
t_1 = the temperature at which the k factor was calculated, °C

The value for θ is not a constant. At a value of 1.072, the reaction rate doubles or halves over a 10°C temperature change. If the value of 1.047 were to be used, the reaction rate would double or halve over a 15°C temperature change. The use of this equation for temperature differences of plus or minus 5°C from the temperature at which the basic k-value was calculated is considered most appropriate.

Example Problem 1.4 shows how to use this equation to calculate the time required for a specific reduction in concentration of reactant to occur based on a specified initial reaction rate constant and a specified temperature.

Example Problem 1.4

Assume a first-order kinetic reaction with a measured k-value of 25 per day at 20°C. Based on a value for θ of 1.072, calculate the k-value at 22°C.

Solution

$$k_{22} = (25/d) \, (1.072)^{(22-20)} = 28.7/d$$

One of the reasons for calculating a reaction rate is to determine the time it would take for a specific reaction to occur. If it is desired to reduce the concentration of a reactant by 100 mg/L, for example, knowing the k-value can determine the time needed for that reaction to occur. Example Problem 1.5 illustrates this concept.

Example Problem 1.5

Given a reaction rate of 30/day, how long will it take to reduce the concentration of a reactant from 118 mg/L to 18 mg/L in a first-order reaction?

Solution

From Equation 1.4, the equation of this reaction is:

$$18 \text{ mg/L} = (118 \text{ mg/L}) \, (e^{-30t})$$
$$\text{Ln} \, (18/118) = -30 \, t$$
$$-1.88 = -30 \, t$$
$$t = 0.06 \text{ d} = 1.5 \text{ hours}$$

Thus, a detention time of 1.5 hours in the reactor should be sufficient to reduce the original concentration to the desired concentration at the given k-value.

1.12   REACTIONS COMMON TO WASTEWATER TREATMENT

1.12.1   OXIDATION-REDUCTION REACTIONS

The terms "oxidation" and "reduction" refer to the addition or removal of electrons to or from an element. The element that gives up the electrons is being oxidized, and is, therefore, the reducing agent, and the element that accepts the electrons (the "electron acceptor") is being reduced and is, therefore, the oxidizing agent. The reducing agent is oxidized and the oxidizing agent is reduced. Oxidation can also mean the gain of oxygen atoms or loss of hydrogen atoms, and reduction can also mean the gain of hydrogen atoms or the loss of oxygen atoms.

The rusting of iron, for example, is an oxidation-reduction reaction because electrons are removed from the ferrous atoms and transferred to the oxygen atoms to form a Fe_2O_3 compound, or ferric oxide. After the electrons are transferred, the oxygen ions and the ferrous ions are held together by the electrostatic forces due to the charges on the ions in the structure of the ferrous oxide. In this case, the iron has lost electrons and the oxygen has gained an equal number. The oxygen, then, is the oxidizing agent and the iron is the reducing agent.

Oxidation and reduction reactions always occur together in a reactor as a result of one or more compounds or elements dissociating in the water and new compounds being created from the residual ions. For example, bisulfite can be used in the removal of excess chlorine (hypochlorite) after that chemical has been used to treat water. The removal of chlorine in this case is a oxidation-reduction reaction as shown in the following text. The sulfur loses two electrons in this process, while the chlorine gains two:

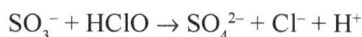

$$SO_3^- + HClO \rightarrow SO_4^{2-} + Cl^- + H^+$$

The oxidation number of an element is equal to the valence of the element. Both the oxidation number and the sign change with the nature of the charge of the ion when formed from the neutral atom. The oxidation number of the chlorine in hydrochloric acid, for example, is -1; in hypochlorous acid the oxidation number of the chlorine is $+1$. The oxidation number of the chlorine in chloric acid ($HClO_3$) is $+5$; in perchloric acid ($HClO_4$) the oxidation number of the chlorine is $+7$.

More detailed information on oxidation–reduction equations can be found in the *Environmental Chemistry* book in this series.

1.12.2 ION-COMBINATION REACTIONS

It is noted that not all reactions that involve ions are oxidation–reduction reactions. Many such reactions are called ion-combination reactions, or sometimes precipitation reactions. These reactions involve no change to the valence (oxidation number) of the reacting chemicals.

Consider, for example, the case of copper sulfate and sodium hydroxide in an aqueous solution. The compounds will react in the following manner. First, the two base compounds will dissociate into their respective ionized forms, as follows:

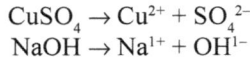

$$CuSO_4 \rightarrow Cu^{2+} + SO_4^{2-}$$
$$NaOH \rightarrow Na^{1+} + OH^{1-}$$

The resulting reactions form $Cu(OH)_2$ and Na_2SO_4 according to the following equation:

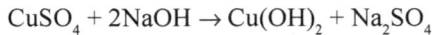

$$CuSO_4 + 2NaOH \rightarrow Cu(OH)_2 + Na_2SO_4$$

All of the reactants in this equation have the same valence on both sides of this equation.

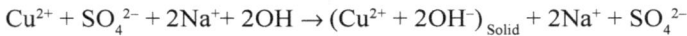

$$Cu^{2+} + SO_4^{2-} + 2Na^+ + 2OH \rightarrow (Cu^{2+} + 2OH^-)_{Solid} + 2Na^+ + SO_4^{2-}$$

This is typical of ion-combination reactions and this type of reaction must not be confused with a true oxidation-reduction reaction. More details on ion-combination reactions can be found in the *Environmental Chemistry* book in this series.

1.12.3 pH AND ALKALINITY

Several of the reactions in wastewater treatment tend to be pH dependent. The pH of a substance is defined as the negative of the logarithm of the hydrogen ion concentration. As a result, although the relationship is not linear, when the hydrogen ion concentration is high, the pH is low and when the hydrogen ion concentration is low, the pH is high. A condition of low pH is considered acidic and a condition of high pH is considered basic, or alkaline.

1.12.4 BUFFERING

An acid condition or an alkaline condition creates a "buffer" in the water. Alkalinity buffers against increasing alkalinity and an alkaline condition buffers against an increasing acidic condition. The strength of a buffer, then, is a measure of the ability of the water to absorb more acid or more alkalinity without causing a significant change in the pH of the water.

Alkalinity is a measure of the ability of the water to neutralize acids, in essence to absorb additional hydrogen ions, without a significant change in the pH of the water. Acidity is a measure of the ability of the water to absorb electron donors without causing a significant change in the pH of the water.

There are generally three forms of alkalinity of importance to wastewater treatment. The form of alkalinity is a function of the pH of the water at the time and the name is reflective of the procedure used to determine the alkalinity in the laboratory.

The three forms of alkalinity of concern are (1) phenolphthalein alkalinity, which is the alkalinity above a pH of 8.3; (2) carbonate alkalinity, which is the alkalinity below a pH of 4.5; and (3) a mix of carbonate and noncarbonate alkalinity, called bicarbonate alkalinity, which exists at a pH between 4.5 and 8.3.

1.12.5 MEASURES OF ALKALINITY

Alkalinity is generally expressed, regardless of the form, in terms of mg/L of calcium carbonate ($CaCO_3$) equivalent. These alkalinity values may be calculated in the laboratory by means of a titration of the water sample with sulfuric acid. One of the principal reasons for expressing alkalinity in this way stems from its definition, which is the algebraic sum of all titratable bases above a pH of about 4.5. In wastewater treatment, these are usually limited to the carbonate species and any free ions of hydrogen or hydroxide. The sum of the hydrogen ions is subtracted from the sum of the carbonate species and the hydroxide ions. Alkalinity is determined through a titration process as the concentration of acid required to lower the pH of water to 4.5. Alkalinity is generally expressed in terms of mg/l of $CaCO_3$.

Example Problem 1.6 shows how to calculate the alkalinity using the results of a sulfuric acid titration procedure. See *Standard Methods for the Examination of Water and Wastewater*, latest edition, for details of the titration method.

Example Problem 1.6

Assume that a 150 mL water sample is titrated with 0.02 N sulfuric acid. If it takes 3.5 mL of acid to reach the phenolic end point and an additional 11.5 mL to reach the mixed bromocresol green-methyl red color change, what are the phenolphthalein and total alkalinities of this solution? Based on those results, determine the carbonate and bicarbonate split within the total alkalinity value.

Solution

Alkalinity is measured as mg/L of calcium carbonate using the following equation:

Alkalinity as mg/L of $CaCO_3$ = (mL of titrant × normality of acid × 50,000)/mL of sample

(The 50,000 come from the atomic weight of the acid and its normality.)

Alkalinity = [(3.5) (0.02) (50,000)]/150 = 23.3 mg/L (as $CaCO_3$)

The total alkalinity = [(15) (0.02) (50,000)]/150

= 100 mg/L (as $CaCO_3$)

Samples containing both carbonate and bicarbonate alkalinity have a pH greater than 8.3. The phenolic endpoint titration represents one half of the carbonate alkalinity. The bicarbonate alkalinity is then the total alkalinity minus the carbonate alkalinity.

Therefore,

Carbonate alkalinity = 2 × 23.3 = 46.6 mg/L as $CaCO_3$

Bicarbonate alkalinity = 100−46.6 = 53.4 mg/L as $CaCO_3$

1.13 COAGULATION AND FLOCCULATION

The chemicals added to wastewater during treatment are selected to do specific tasks. One of the most important of those tasks is to convert colloidal particles, chemical components, and particles of waste that are suspended in solution such that they will not settle out into particles that are big enough and heavy enough to settle out of the water in the sedimentation basins. Many of the dissolved and fine suspended particles are useful in the biological process of treatment, but many others are either harmful to the biology or of no use to it and therefore pass through the system untreated unless chemical reactions are created to assist with removal.

Removal of these particles and substances fundamentally requires converting these very tiny suspended particles and dissolved materials into large enough suspended particles that they will settle out in a clarifier or sedimentation basin. It is also desirable that all of that happens fairly quickly to reduce treatment time and therefore reactor volume requirements.

Physically, there are two problems here. The first is the removal of hydrophilic particles (those that are happy living in the presence of water) and the second is the removal of hydrophobic particles (those that are not happy living in the presence of water). In addition, they all have such a high surface area to mass that they will simply not settle if left untreated.

In nature, and in the wastewater, these particles are kept apart from each other by repulsive electrical charges. These repulsive forces work much the way magnets work when poles of the same charge are placed near each other. When the appropriate chemicals are added, the repulsion forces are overcome by a suppression action on the external electrical charges and the particles begin to come together. Other chemicals, called polymers, create "strands" of chemical to which the colloidal particles attach until the strand becomes heavy enough to settle. When a sufficient number of charges has been adequately suppressed, or the strands have become long enough and heavy enough, the particles gain enough mass to settle out of the water in the sedimentation basins. See Chapter 3 for further discussion of the sedimentation processes at work in wastewater treatment.

Chemically, the concept of converting the submicron particles to suspended matter (or, in essence, the destabilization of the particles by suppression of the charged surface layers) is called coagulation, and the aggregation of the destabilized particles into a large enough mass to settle is called flocculation. In wastewater treatment, the terms are usually used together, as in "coagulation/flocculation" or they are referred to collectively as "coagulation." These two steps must be followed by a sedimentation step if the coagulated particles are to be removed from the wastewater. Therefore, it is also common to add this third step to the chain and refer to the entire process as "coagulation/flocculation/sedimentation."

Coagulation is generally a very rapid process, while flocculation tends to be a much slower process. It is helpful, then, to ensure that the chemicals come in contact with the colloids as quickly as possible. This is usually done with a rapid mixer, often an in-line mixer, which very violently mixes the chemical into the water in somewhere between 30 and 60 seconds of contact time. The mixture then goes to the flocculation basin where the chemicals and wastewater are slowly and gently stirred. This slow, gentle stirring allows the chemicals to suppress the electrical charges and then cause the particles to bump into each other and attach together without adding so much energy to the system that the combined particles are then ripped apart again. This step typically requires about 30 minutes for completion. At the end of the flocculation period, the water enters a quiet settling zone in a sedimentation basin where it sits for 2 to 4 hours to allow the coagulated and flocculated masses to settle out.

The settled matter is removed mechanically from the bottom of the reactor and handled as waste sludge.

These three separate steps need to occur sequentially. The longer the flocculation and sedimentation steps, the better the overall colloidal removal rate, up to a point. There is a point of diminishing returns from longer detention times and those are close to the time limits noted earlier. It is also important to note that this process does not generally remove all the colloidal particles, so some form of filtration is also necessary to effectively remove the rest. Filtration could be done without prior sedimentation too, but it very rapidly clogs the filter and creates major maintenance problems.

1.14   HARDNESS OF WATER

Discussions of alkalinity lead to a discussion of hardness in water. Hardness is generally defined as the sum of the calcium (Ca) and magnesium (Mg) ions in the water, expressed in terms of mg/l of concentration. Hardness is also expressed, however, in terms of mg/L of $CaCO_3$ equivalence, in the same way that alkalinity is expressed.

Both hardness and alkalinity are discussed in more detail later in this volume as the processes in which they are important are addressed.

1.15   CHEMICAL OXYGEN DEMAND

1.15.1   CONCEPTS

The Chemical Oxygen Demand (COD) of a wastewater is the total amount of oxygen needed to chemically oxidize all of the organics in the wastewater to carbon dioxide and water. It is measured in mg/L. The method for determining the COD is a reflux (vaporizing and then condensing) reaction typically involving potassium dichromate, sulfuric acid, and silver sulfate. See *Standard Methods* for proper test procedures. This is a relatively rapid test procedure that gives reasonably consistent results.

1.15.2   RELEVANCE

The COD test is used to test the organic strength of a wastewater because it is much faster (hours) than the standard Biological Oxygen

Demand (BOD) test (see Chapter 2), which takes five days to complete. Good correlations can generally be developed over time between the COD and BOD values for a given wastewater treatment plant, but these correlations depend on the "normal" constituents of the wastewater and will not be consistent among different plants. The correlations may also change over time and they should be verified at each plant on a regular basis.

1.16   TOTAL ORGANIC CARBON

1.16.1   CONCEPTS

Total Organic Carbon (TOC) is also used as a measure of the organic strength of wastewater. This term incorporates a more complex array of compounds. The total carbon concentration includes

- Total Inorganic Carbon—which is the total carbonate, bicarbonate, and carbon dioxide fractions;
- TOC—which includes all carbon atoms covalently bound to organic molecules;
- Dissolved Organic Carbon (DOC)—which is defined as the fraction of TOC that passes through a 0.45 μm filter paper;
- Suspended Organic Carbon—which is the fraction of the TOC that is retained on 0.45 μm filter paper.

The test for TOC involves digesting all of the inorganic carbon to CO_2 at a pH of 2.0 or less and then purging the CO_2 out of the system with an inert gas. The procedures for doing all of that vary. See *Standard Methods* for the appropriate procedures. The incorporation of in-line TOC measurement has been found to be a useful process management tool that is worthy of consideration.

1.16.2   RELEVANCE

The TOC test is used primarily when water reclamation is anticipated or practiced. There are various health issues raised by the quantities of organic and inorganic carbon that can pass through a wastewater treatment plant untreated and enter a receiving water. If that discharge is then used for water supply augmentation, the health concerns are magnified. A TOC test can indicate the relative magnitude of the threat and the need for further carbon reduction in the wastewater treatment plant effluent, or a reduction in the relative volumes used for water supply augmentation or other wastewater reuse options. This is a surrogate, or indirect, test method when used for these purposes, which indicates the general water quality, relative to residual carbon concentrations.

1.17   FATS, OIL, AND GREASE

1.17.1   CONCEPTS

Fats, Oil, and Grease, more commonly referred to as "FOG," consist of a variety of organic substances, including hydrocarbons, animal fat and grease, oils, waxes, and various fatty acids that generally originate from households, food preparation operations, and restaurant operations. They do not typically break down well in wastewater treatment plants, but they can be made to float in sedimentation basins and air floatation units for relatively easy removal from the wastewater being treated. The collected materials are solid or semisolid in nature and are generally combined with other waste solids or sludges for treatment prior to disposal.

1.17.2   RELEVANCE

The issue with FOG components is their natural tendency to adhere to the walls of pipes and pump stations, significantly reducing the carrying capacity of the pipes, particularly in colder weather, and interfering with the proper operation of pumps and floats in the pumping stations. FOG that carries over to the treatment plant can also interfere with the operation of flow measuring devices, main pump station equipment, and sedimentation operations throughout the plant. Although FOG components are organic in nature, they are resistant to the biological treatment most commonly provided in wastewater treatment plants because of the short resident times in the various treatment units, compared with the long retention times needed for biological FOG degradation.

The test for FOG concentrations involves an extraction procedure using n-hexane and gravimetry to determine the volume of extracted components. It is noted that FOG components often adhere to the sampling and testing equipment so care is needed to ensure reliable results. See *Standard Methods* for the proper testing techniques for FOG.

1.18 MATERIAL BALANCE CALCULATIONS

A material balance calculation is the application of the law of conservation of mass, which states that mass can be neither created nor destroyed. Mass can, of course, be converted to different forms, but it must all still be there at the end of the calculations. It is fundamentally a process of accounting for what happens to materials during a chemical reaction or treatment process. The basic equation for mass balance is:

$$\text{Input} = \text{Output}$$

This is an oversimplification relative to wastewater treatment since the components are often converted to other things during the treatment process. This can make direct measurements of the input components difficult. The more complete equation is the following:

$$\text{Input} + \text{Generation} - \text{Output} - \text{Consumption} = \text{Accumulation} \quad (1.10)$$

The "input" components are those that enter the system through the system boundaries, such as the components of wastewater entering a reactor through a pipe. The "generation" components are those that are produced inside the system through combinations with other constituents in the wastewater. The components of "output" are those fractions that leave the system, such as over the effluent weir in a treatment plant. The components of "consumption" are used inside the reactor as building blocks for new compounds created by the reactions inside the reactor. And the remaining fractions stay in the reactor, but are not used in any reaction, so they accumulate inside the reactor.

The calculation of mass balances requires an initial assumption that the process being measured is in a "steady-state" condition. This means that the inflow and the outflow volumes are the same, the concentration of the components of the inflow are not changing, and the reactions that are occurring inside the reactor or system are also continuing at the same constant rate.

Material balance problems generally involve a description of the process being measured, the values of several process parameters or variables that are known, and a list of the values to be determined. The solution then follows three key steps, as follows:

1. A flow chart of the process being evaluated is drawn and labeled. This is referred to as a "block diagram." The values of known parameters are shown and symbols are used to indicate the value of unknown variables.
2. A basis of calculation is then selected. This is usually a concentration or flow rate consistent with one of the known values.

3. A material balance equation(s) is then written to describe the situation being evaluated using the fewest unknown variables possible. The maximum number of independent equations that can be written for each system is equal to the number of species in the input and output streams of the system.
4. The equations derived in step 3 are then solved for the unknown quantities to be determined.

A simple example is the following Example Problem 1.7.

Example Problem 1.7

Given a wastewater treatment plant discharge of 1.5 million gallons per day (mgd) with a dissolved oxygen concentration of 2.5 mg/L and a receiving water stream flowing at 52.5 cubic feet per second (cfs) with a dissolved oxygen concentration of 7.9 mg/L, what is the resulting concentration of oxygen in the stream when complete mixing of the two flows has occurred?

Solution

There are several steps to this solution. The first is to be sure that all the same types of parameters are measured in the same units of measure. Flow, for example, is given as mgd for the treatment plant discharge, but in cfs for the stream. The discharge from either one has to be converted to the units of measure for the other. It is not important which units are converted, but they all have to be the same in the end. In this case, the flow from the treatment plant is converted to cfs, as follows:

$$(1.5 \times 10^6 \text{ gallons/day}) (1 \text{ day/24 hours}) (1 \text{ cf/7.48 gallons})$$
$$(1 \text{ hour/3600 seconds}) = 2.3 \text{ cfs}$$

The boundaries of the system to be measured are then established such that all of the components are within the boundary. Here the boundaries are the stream from the point of input from the treatment plant to the point of complete mixing downstream. Those system boundaries can be represented by a block diagram, as follows:

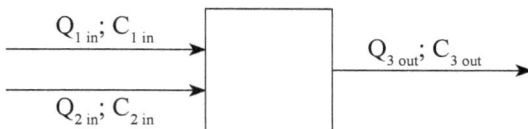

Where:

$$Q_{1\,in} = 2.3 \text{ cfs} \qquad C_{1\,in} = 2.5 \text{ mg/L}$$
$$Q_{2\,in} = 52.5 \text{ cfs} \qquad C_{2\,in} = 7.9 \text{ mg/L}$$
$$Q_{3\,out} = \text{Unknown} \qquad C_{3\,out} = \text{Unknown}$$

If the system is at steady state then the total flow in must equal the total flow out:

$$Q_{3\,out} = Q_{1\,in} + Q_{2\,in} = 2.3 \text{ cfs} + 52.5 \text{ cfs} = 54.8 \text{ cfs}$$

The resulting concentration of dissolved oxygen, which is the ultimate calculation desired, is calculated from the following equation:

$$\text{(The sum of the input flows} \times \text{their concentrations)} =$$
$$\text{(The sum of the output flows} \times \text{their concentrations)} \qquad (1.11)$$

Or:

$$(2.3 \text{ cfs}) (2.5 \text{ mg/L}) + (52.5 \text{ cfs}) (7.9 \text{ mg/L}) = (54.8 \text{ cfs}) (C_{3\,out})$$
$$C_{3\,out} = 7.67 \text{ mg/L}$$

This same material balance approach is equally applicable to nutrient concentrations, such as nitrogen and phosphorous. These nutrient issues arise seasonally, in most cases, but are very important to water quality determinations downstream of point and nonpoint discharges.

1.19 EMERGING CHEMICALS OF CONCERN

Wastewater may also contain a variety of chemicals and compounds that have not been mentioned here, but which can cause significant disruption to the effective treatment of the wastewater. Table 1.4 provides a partial list of the most common of those emerging chemicals of concern, along with source or chemical type. Pharmaceuticals and Personal Care Products, such as hair sprays and fragrances, collectively known as "PPCPs," have been seen as the leading sources of these compounds, but other sources are emerging as prime candidates as well. Table 1.4 indicates that pesticides, detergents, fire retardants, insect repellants, hydrocarbon spills and releases (such as leaking underground storage tanks for fuels and fuel oils), and plasticizers are also prime suspects as sources. Currently, there are no good options available for dealing with these chemicals and compounds. This is an emerging area of interest in wastewater treatment

Table 1.4. Emerging chemicals of concern

Chemical measured	Type of chemical or use
3-beta-coprostanol	Steroid
4-nonylphenol	Detergent metabolite
4-tert-octylphenol	Detergent metabolite
Anthracene	Polycyclic aromatic hydrocarbons
Beta-sitosterol	Steroid
Bisphenol A	Fire retardant
Caffeine	Stimulant
Carbamazepine	Antiepileptic drug
Cholesterol	Steroid
Diazinon	Pesticide
Diethylhexyl phthalate	Plasticizer
Diphenhydramine	Antihistamine drug
D-limonene	Fragrance compound
Estrogenic steroids	Steroids
Fluoranthene	Polycyclic aromatic hydrocarbons
Fluoxetine	Antidepressant drug
Galaxolide (HHCB)	Fragrance compound
Indole	Fragrance compound
N,N-diethyltoluamide	Insect repellant
Nonylphenol, dithoxy—total	Detergent metabolite
Nonylphenol, monoethoxy—total	Detergent metabolite
Para-cresol	Preservative
Para-nonylphenol—total	Detergent metabolite
Phenanthrene	Polycyclic aromatic hydrocarbons
Phenol	Disinfectant chemical
Phenytoin	Antiepileptic drug
Pyrene	Polycyclic aromatic hydrocarbons
Selective serotonin uptake inhibitors (SSRIs)	Antidepressant drugs
Skatol	Fecal indicator
Stigmastanol	Steroid
Tonalide (AHTN)	Fragrance compound
Tri(2-chloroethyl)phosphate	Fire retardant
Triclosan	Disinfectant chemical
Valproate	Antiepileptic drug

research. A significant amount of research needs to be done before wastewater treatment plants can be expected to effectively manage these types of compounds on a regular basis.

BIBLIOGRAPHY

Wastewater Technology Fact Sheet Dechlorination. 2014. Retrieved from EPA.gov: http://water.epa.gov/scitech/wastetech/upload/2002_06_28_mtb_dechlorination.pdf, (June 5, 2014).

Greywater Reuse. 2014. *Greywater Action for a Sustainable Water Culture*, Retrieved from greywateraction.org: http://greywateraction.org/content/about-greywater-reuse, (April 21, 2014).

Anonymous. 2012. *faculty.kfupm.edu.sa*, Retrieved from Chapter 4 Material Balances and Applications: http://faculty.kfupm.edu.sa/CHE/aljuhani/New_Folder/Material%20%20balance.pdf, (2012, May 12).

Barnes, K.K. 2002. *Pharmaceuticals, Hormones, and Other Organic Wastewater Contaminants in U. S. Streams 1999-2000*. Washington, DC: USGS.

Davis, M.L. 2011. *Water and Wastewater Engineering: Design Principles and Practice*. New York, NY: McGraw Hill Book Co.

Felder, R.M. 2000. *Elementary Principles of Chemical Processes*. 3rd Ed. New York, NY: John Wiley & Sons.

Hammer, M.J. 2008. *Water and Wastewater Technology*. Upper Saddle River, NJ: Pearson Prentice Hall.

Hill, J.W. 1996. *General Chemistry*. Upper Saddle River, NJ: Prentice Hall.

Kolpin, D.W. 2002. "Pharmaceuticals, Hormones, and Other Organic Wastewater Contaminants in U.S. Streams, 1999−2000: A National Reconnaissance." *Environmental Science and Technology* 36, no.6, pp. 1202–1211. doi: http://dx.doi.org/10.1021/es025709f.

Metcalf & Eddy/AECOM. 2014. *Wastewater Engineering Treatment and Resource Recovery*. New York, NY: McGraw-Hill Publishers.

Rice, E.W. 2012. *Standard Methods for the Examination of Water and Wastewater*. Washington, DC: American Water Works Association/American Public Works Association/Water Environment Federation.

Richardson, S.D. 2003. "Disinfection by-Products and Other Emerging Contaminants in Drinking Water." *TrAC Trends in Analytical Chemistry* 22, no. 10, pp. 666–684.

Sawyer, C.N. 1994. *Chemistry for Enviromental Engineering*. New York, NY: KMcGraw Hill Book Co.

Yen, T.F. 1999. *Environmental Chemistry, Essentials of Chemistry for Enginering Practice*. Upper Saddle River, NJ: Prentice Hall PTR.

CHAPTER 2

Biology Considerations

2.1 INTRODUCTION

Almost all domestic wastewater treatment is biological. There are some physical treatment components that are included in the overall system, but they are primarily there to remove the biological mass that has grown within the system using the wastewater components as nutrients to support that growth. Without those physical components none of the biological components would be able to continue to function and the contaminants in the wastewater would continue to recycle within the system.

The basic concept of domestic wastewater treatment is to convert as much of the suspended and dissolved contaminants in the incoming wastewater to single cell organisms and to then remove those organisms from the waste stream, thereby also removing the contaminants from the waste stream. Consequently, there is a serious need to understand some of the basic biology of wastewater treatment in order to be able to understand the dynamics of the treatment process.

The following six categories or types of organisms need to be considered in wastewater treatment:

Bacteria
Viruses
Algae
Fungi
Protozoans
Microscopic multicellular organisms

2.2 BACTERIA

Bacteria are microscopic prokaryotic organisms that can feed by selective intake of nutrients dissolved in water and that divide into two equal cells by binary fission. Bacteria come in a large variety of sizes, shapes, types, and forms. Not all types and forms are relevant in most wastewater treatment plant operations but the following types are important.

2.2.1 HETEROTROPHIC BACTERIA

Heterotrophic bacteria use organic matter as an energy source and as a carbon source for synthesis. These components come from the constituents of the wastewater. Organic contaminants compose the bulk of the contaminants in most domestic wastewater; therefore, heterotrophic bacteria tend to dominate in most wastewater treatment plants. All of the organic matter is degraded by heterotrophs, usually aerobic heterotrophs, except in anaerobic digesters, or in anaerobic lagoons specifically designed to operate under anaerobic conditions.

2.2.2 AUTOTROPHIC BACTERIA

Autotrophic bacteria use inorganic substances for energy and carbon dioxide as a carbon source for synthesis. These forms of bacteria tend to dominate in wastewater dominated by certain industrial flows, rather than in domestic wastewater. They are an important component in wastewater, however, and may not be ignored. Certain inorganic matter is best degraded by autotrophs, such as nitrifying bacteria (which are important in the removal of nitrogen), sulfur reducing bacteria, and iron reducing bacteria, all of which are anaerobic bacteria.

2.2.3 AEROBIC BACTERIA

Aerobic bacteria, or aerobes, require oxygen for synthesis, much the way people do. The bacteria utilize oxygen that is tied up in organic matter when they can, but they also utilize excess dissolved oxygen (DO) directly from the wastewater during synthesis. Wastewater with a sufficient concentration of DO will operate as an aerobic system, which indicates that aerobes dominate the bacteria mass, while wastewater deficient in DO will operate in an anaerobic mode. Anaerobic operation generally results in unacceptable odor problems and is not used except in completely closed portions of the treatment process for that reason. Maintaining sufficient DO in the wastewater for effective aerobic treatment is an essential element of treatment plant operation.

2.2.4 ANAEROBIC BACTERIA

Anaerobic bacteria, or anaerobes, use nitrogen, sulfur, or iron (generally in that order of preference) for synthesis instead of oxygen. They also generate hydrogen sulfide among other objectionable gaseous by-products of organic degradation. Anaerobic degradation of organic matter tends to be faster than aerobic degradation, but the objectionable by-products are great enough to discourage their use. Except with respect to phosphorus removal, sludge digestion, or both, the extra time required for aerobic degradation is not sufficient to warrant suffering the odor problems associated with anaerobic degradation.

2.2.5 FACULTATIVE BACTERIA

Facultative organisms can live in either aerobic or anaerobic environments, but usually favor one or the other. Reference will be made to facultative aerobes when referring to facultative organisms that prefer aerobic conditions, but can function in anaerobic environments. Facultative anaerobes prefer anaerobic conditions, but can also function well in aerobic environments.

2.3 VIRUSES

Viruses are obligate intracellular parasites that replicate only within a living host cell. The viruses of concern are generally disease causing and are associated with the discharge of animal fecal matter. Certain animal feces also contain viruses that are transmittable to humans through inadequate watershed management or inadequate water treatment and physical contact. See Table 2.1 for various viruses of concern and the diseases associated with them.

2.4 ALGAE

Algae are microscopic photosynthetic plants having no roots, stems, or leaves. They are generally autotrophic and they produce oxygen as a by-product of synthesis. The generation of oxygen is convenient in wastewater treatment, but algae forms tend to be long and filamentous and they tend to clog the discharge weirs of treatment plants. They also foul filters, clog pipes, and otherwise interfere with proper treatment plant operations.

Table 2.1. Pathogens often excreted by, or ingested by, humans

Group or name	Type of organism	Associated diseases
Acanthamoeba	Bacterium	Eye infections
Adenoviruses	Virus	Respiratory and eye infections
Ancylostoma duodenale	Helminth	Hookworm
Ascaris lumbricoides	Helminth	Roundworm
Caliciviruses	Virus	Diarrhea
Campylobacter jejuni	Bacterium	Campylobacteriosis
Chlamydia trachomatis	Bacterium	Eye disease and blindness
Coxsackie viruses	Virus	Aseptic meningitis, herpangina, myocarditis
Cryptosporidium	Parasite (protozoan)	Diarrhea
Echoviruses	Virus	Aseptic meningitis, diarrhea, respiratory infections
Entamoeba histolytica	Parasite (protozoan)	Amoebic dysentery
Enterobius vermicularis	Helminth	Pinworm
Escherichia coli	Bacterium	Diarrhea, abdominal cramping, and pain
Fasciola hepatica	Bacterium	Liver disease
Giardia lamblia	Protozoan	Diarrhea
Hepatitis A virus	Virus	Infectious hepatitis
Hepatitis E virus (rare in the U.S.)	Virus	Liver disease and cancer of the liver
Hymenolepis nana	Helminth	Dwarf tapeworm
Legionella pneumophila	Bacterium	Fever, chills, pneumonia, anorexia, muscle aches, diarrhea, and vomiting
Misc. *Salmonellas*	Bacterium	Gastroenteritis
Misc. *Vibrios*	Bacterium	Diarrhea

(Continued)

Table 2.1. (*Continued*)

Group or name	Type of organism	Associated diseases
Misc. viruses	Virus	Gastroenteritis, diarrhea
Necator americanus	Helminth	Hookworm
Noroviruses	Virus	Gastroenteritis
Polioviruses	Virus	Aseptic meningitis, poliomyelitis
Rotavirus	Virus	Gastroenteritis
Salmonella paratyphi	Bacterium	Paratyphoid fever
Salmonella typhi	Bacterium	Typhoid fever
Shigella	Bacterium	Dysentery (bacillary)
Strongyloides stercoralis	Helminth	Threadworm
Trichinella	Parasite (Protozoan)	Trichinosis (also called Trichinellosis)
Trichuris trichiura	Helminth	Whipworm
Vibrio cholerae	Bacterium	Cholera
Yersinia enterocolitica	Bacterium	Gastroenteritis

Algae often indicate a nutrient limiting situation in that a sudden excess of a certain nutrient will cause an algal bloom, or sudden spurt of algae growth in a reactor or receiving water.

Blooms caused by nitrogen, for example, are said to occur in a nitrogen-limited environment to which significant concentrations of available nitrogen are suddenly added. This is the primary reason for algae blooms in lakes and ponds in the summer where residential properties line the shore and wastewater is disposed through subsurface disposal fields located close to the shore. The wastewater discharged is high in nitrogen compounds that flow through the soil to the body of water and then provide vast quantities of available nitrogen for the algae. Nonpoint source runoff from lawns, golf courses, farms, and other areas tends to be a larger source of nutrient addition to rivers and streams than subsurface disposal.

Similarly, phosphorous—which is of more concern today—can be discharged to the soil through wastewater or, more commonly, through fertilizers used for grass and plants throughout the watershed. When the excess phosphorous reaches the receiving water, an algae bloom often results. In general, fresh water environments are going to be phosphorous limited, whereas marine environments will tend to be nitrogen limited.

2.5 FUNGI

Fungi are eukaryotic aerobic microbes that are nonphotosynthetic and include diverse forms such us unicellular yeasts and filamentous molds. Since they lack chlorophyll, fungi have to get nutrition from organic substances, and many of them obtain their food from dead organic matter. Fungi grow best in low pH conditions that are high in sugars.

In this context, "sugar" refers to a carbohydrate product of photosynthesis. These are typically long-chain compounds of hydrogen, oxygen, and carbon and they will generally contain one or more chemical groups called "saccharose" groups. A "simple sugar," which is one composed of chains containing two to seven carbon atoms, are referred to as "monosaccharide sugars." These are things like glucose (or dextrose) and fructose (or levulose). Sucrose is a disaccharide generally originating as cane or beet sugars; lactose, a milk-based sugar; and cellobiose, a sugar originating from cellulose.

Fungi tend to create problems in treatment plants because they tend not to settle well and to cause bulking of the sludge in sedimentation basins.

2.6 PROTOZOANS

Protozoans (meaning "first animals") are a large group of unicellular eukaryotic organisms that belong to the kingdom Protista. Protozoans get their name because they have the same type of feeding as animals, that is, they are heterotrophic and obtain cellular energy by metabolizing organic matter. Protozoans are mostly aerobic or facultative with regards to oxygen requirements. The most commonly associated with wastewater include amoeba, flagellates, free-swimming ciliates, and stalked ciliates.

2.7 MICROSCOPIC MULTICELLULAR ORGANISMS

Microscopic multicellular organisms are generally heterotrophic aerobes that act as control organisms keeping the other lower order organisms growing in proper proportions in the treatment plant. This group includes rotifers, helminthes, and microscopic crustaceans. Rotifers are associated with cleaner waters and are normally found in well-operated wastewater plants. Helminthes have a special capability to encyst when living conditions are unfavorable, such as having been excreted from a host organism into a watershed, and then un-encysting when conditions are suitable again, such as after having been ingested by a warm-blooded animal, including humans. Microscopic crustaceans such as Daphnia and Cyclops are two examples of interest to wastewater operators.

2.8   PATHOGENS

Many of the organisms identified earlier are also pathogenic to humans. This means that they generally cause adverse health effects in humans when ingested. Table 2.1 lists typical pathogens excreted by humans or often ingested by humans through inadequately treated drinking water, listed by organism and the disease associated with that organism.

2.9   INDICATOR ORGANISMS

It is very difficult to measure specific species of organism that are harmful, but not so difficult to measure specific species that are always present when the harmful ones are present and always absent when the harmful ones are absent. These measurable organisms are called "indicator organisms" in that they indicate, with very high reliability, the presence or absence of pathogenic organisms. The most common of these indicator organisms is coliform bacteria. Two forms of coliform are typically measured in wastewater treatment: total coliform and fecal coliform. The fecal coliform count measures coliform bacteria typically found in the feces of humans, while total coliform indicate the presence of pathogenic organisms from other sources. Of late, a third bacterium, *Enterococcus*, a genus of lactic acid bacteria of the phylum Firmicutes, is also being included in certain discharge permits. Enterococci, Gram-positive cocci that often occur in pairs (diplococci) or short chains, are difficult to distinguish from streptococci on physical characteristics alone. Two species are commonly found in the intestines of humans: *Enterococcus faecalis* (90 to 95 percent) and *Enterococcus faecium* (5 to 10 percent).

The concentration (most probable number, MPN) of coliform is typically measured using one of two techniques: a multiple-tube fermentation technique or a membrane filtration technique using m-ENDO for total coliform and m-FC for fecal coliform. Both are incubated for 24 hours, plus or minus 2 hours. The total coliform are incubated at 35°C, plus or minus 0.5°C. Fecal coliform are incubated at 44.5°C plus or minus 0.5°C. Both techniques are fully described in Standard Methods. The membrane filtration technique is much simpler to perform, but does require some care during implementation to avoid inadvertent contamination of the petri dishes being used. It is also a quantitative technique, however, which has significant advantages over the qualitative multiple-tube technique.

The MPN concentration of *Enterococcus* bacteria is also measured using a filtration technique, (EPA Method 1600) but using m-EI agar plates for incubation. Incubation is done for 24 hours, plus or minus 2 hours, at a temperature of 41°C, plus or minus 0.5°C.

2.10   BIOLOGICAL OXYGEN DEMAND

BOD stands for Biological Oxygen Demand and sometimes for Biochemical Oxygen Demand, depending on the source. Both terms mean the same thing: the amount of oxygen needed to biologically degrade (oxidize) all of the organic matter present in a wastewater sample. It is the most commonly used parameter to define the relative strength of domestic (and most industrial) wastewater. It is measured in terms of mg/L of DO. Those units are used because they are consistent with the measurement of the organic content of the wastewater, which is also calculated in mg/L.

BOD is determined from a BOD test. The standard test procedures are fully defined in Standard Methods. Although this test is one of the oldest and most commonly used tests in wastewater treatment, it is also one of the most difficult to reproduce with any reliable consistency. In short, BOD bottles, most commonly 300 mL bottles, but occasionally 60 mL bottles, are filled with a mixture of wastewater sample, growth medium, and sterile, but not distilled, water. The DO concentration is measured and recorded and the bottles are placed in a temperature controlled incubator at 20°C. At the end of 5 or 7 days, the samples are removed from the incubator and the DO content is again measured and recorded. The difference between the initial DO and the final DO is the mass of oxygen used by the volume of actual wastewater in the sample during the incubation period.

Difficulties arise from the mixing of the growth medium and assurance of sterility in that mix, measurement of the DO at the start and at the end of the incubation period with consistent reliability, and from minor variations in the measurement of the actual volumes of wastewater added to each sample bottle. If everything is not extremely carefully done, and no sample or growth medium contamination is allowed to occur, results should be consistent regardless of the dilution factor used in the various sample bottles. The ideal and reality are seldom well aligned with this test.

2.11   BOD FORMULAS OF CONCERN

There are several basic equations relating BOD exerted at any time, t (called BOD_t), the BOD remaining at time, t (called y) and the ultimate BOD, which is also equal to the initial BOD (called L). Those equations are the following:

$$BOD_t = L - y \tag{2.1}$$

$$L = BOD_t + y \tag{2.2}$$

$$y = L - BOD_t \tag{2.3}$$

There are three additional BOD formulae to be concerned with:

$$BOD_t = L\,(1 - e^{-kt}) : \quad \text{BOD exerted in time t} \tag{2.4}$$

$$y = Le^{-kt} : \quad\quad\quad \text{BOD remaining at time t} \tag{2.5}$$

$$L = BOD_t + y : \quad\quad \text{Initial or ultimate BOD} \tag{2.6}$$

Where:

BOD_t = BOD exerted in time, t, in mg/L
y = BOD remaining at time, t, in mg/L
L= Initial or ultimate BOD, in mg/L
k = Reaction rate constant, in 1/day
e = Mathematical e

(Note: Initial BOD and ultimate BOD must be equal values)

2.12   BIOLOGICAL DECAY RATE—k

Biological reactions tend to occur as first-order reactions in which the reaction rate—the rate at which the organic fractions are oxidized—depends on the concentration of the organics and, to a certain degree, the concentration of suitable organisms. The rate is measured in terms of days^{-1}, or 1/day, and indicates the approximate percentage of the remaining organics that will be degraded each day. Since these values are a percentage of the remaining concentrations, not the initial concentrations, it is clearly not possible to achieve complete oxidation within any reasonable time frame. By around day 7 or 8, however, essentially all of the oxidizable organics have generally been oxidized.

The BOD test is also used to determine the kinetic growth rate of the bacteria in the wastewater, the k-rate, as a measure of the time required for complete oxidation of the waste contaminants. For this purpose, up to seven complete sets of BOD bottles are filled and incubated simultaneously. At approximately the same time each day (and the specific time of testing each day is important to record here), one set of bottles is opened and tested for DO concentration. The change in concentration each day is recorded and plotted on a concentration over time graph.

The calculation of BOD exerted as a function of the DO utilized is the following. The equation to use depends upon whether the sample was seeded with bacteria. If seeding was done, the DO utilization of the seed water must be subtracted from the BOD of the wastewater to yield an accurate BOD exerted value.

For seeded wastewater, the equation is the following:

$$BOD_t = \{(D_1 - D_2) - [(S_1 - S_2) \, f]\}/P \qquad (2.7)$$

Where:

BOD_t = BOD exerted in time, t, in mg/L
D_1 = Initial DO in wastewater sample, in mg/L
D_2 = DO in wastewater sample at time, t, in mg/L
S_1 = Initial DO in seed control, in mg/L
S_2 = DO in seed control at time, t, in mg/L
f = Ratio of seeded dilution water volume in sample to volume of seeded dilution water in seed control, unitless
P = Ratio of wastewater volume to total liquid volume in the BOD bottle, unitless

For unseeded wastewater, the equation is the following:

$$BOD_t = (D_1 - D_2) \, /P \qquad (2.8)$$

where the terms are the same as stated earlier.

Once the BOD exerted has been standardized to mg/L, those data are plotted on a graph to verify that they follow a standard BOD curve reasonably well. If that does not happen, the data need to be re-examined since it is likely that the curve will not yield a reliable k-rate value in most cases.

There are many ways that have been developed to determine the k-rate from the observed BOD data over time. They generally do not give consistent results. It is important, therefore, to understand which method is being used and to use the same method consistently. The actual rate is

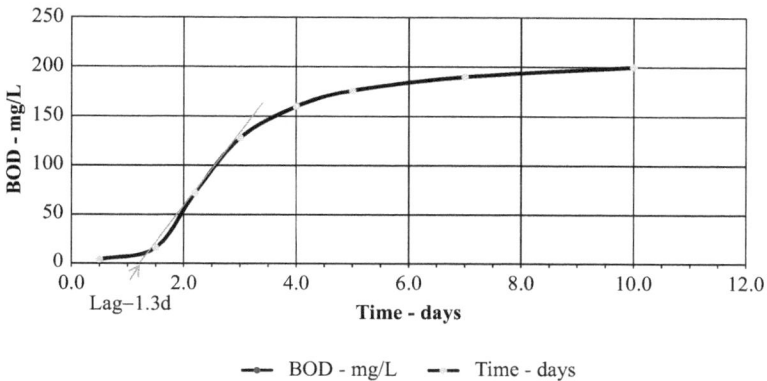

Figure 2.1. Typical BOD curve.

then observed and corrections to the calculations can be made, or an alternative method selected, for future evaluations.

Methods utilized to determine the k-rate include

daily-difference method;
Fujimoto method;
least squares method;
respirometer method;
Thomas method.

The daily difference method and the Thomas method are the methods historically utilized. With the Thomas method, once the curve of BOD exerted over time has been verified, any lag in the start of the BOD exertion is noted. This is determined by drawing a tangent to the rising leg of the BOD curve at the point where the curve begins to break back to the right and the rate of increase starts to decline (see Figure 2.1). The lag, in days, is determined from the point on the x-axis that the tangent line intersects with that axis.

All the BOD time data are then adjusted by the value of the lag. This is an important step in the accuracy of the k-rate calculated from those data because ignoring it lets the lag time extend the time for the reactions to occur and that causes a decrease in the k-rate value below the true value. That results in over-sizing of tanks and extending aeration periods longer than necessary. Not all BOD curves will have a lag.

A new plot of the BOD exertion curve using the new time data, if there is a lag, and the original BOD exerted data is then drawn. This curve

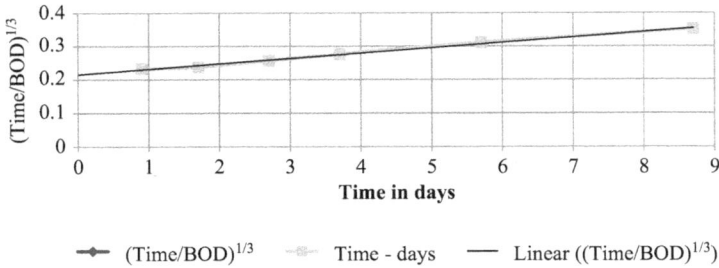

Figure 2.2. $(\text{Time/BOD})^{1/3}$.

plots the following expression on the y-axis against the adjusted time on the x-axis in days:

$$Y = [\text{New time, in days/BOD Exerted, in mg/L}]^{1/3} \qquad (2.9)$$

This plot should yield data points through which a reasonably straight line of best fit can be drawn. The value of k is then determined by the following equation (see Figure 2.2).

If there is no lag in the BOD curve, then the curve shown earlier uses the $(\text{Actual Time/BOD})^{1/3}$ over time to determine the k-rate. That value is calculated by:

$$k = 2.61 \ (B/A) \qquad (2.10)$$

Where:

A = y-axis intercept of the line of best fit
B = Slope of the line of best fit

If there is a lag in the BOD curve, as is the case with the data provided earlier, a more complicated solution is required, as shown by Example Problem 2.1. The values used are taken from Figures 2.1 and 2.2, as appropriate.

Example Problem 2.1

Given the following measured values for BOD from a specific wastewater over time, determine the ultimate BOD concentration (L) and the k-rate value for this waste.

Time, days	0.5	1.0	2.0	3.0	4.0	5.0	7.0	10.0
BOD, mg/L	4	16	72	128	160	176	190	200

Solution

First, the data need to be plotted in a standard BOD curve, as has been done in Figure 2.1. Then a table needs to be constructed showing the time, the corrected time, based on a lag of 1.3 days, as shown on Figure 2.1, the original BOD, and the value of $(Time/BOD)^{1/3}$. That table is shown as follows and the $(Corrected\ Time/BOD)^{1/3}$ is plotted, as in Figure 2.2.

Time, in days	Corrected time, in days	BOD, in mg/L	$(Time/BOD)^{1/3}$
0.5	—	4	—
1.0	—	16	—
2.0	0.7	72	0.2138
3.0	1.7	128	0.2372
4.0	2.7	160	0.2568
5.0	3.7	176	0.2763
7.0	5.7	190	0.3111
10.0	8.7	200	0.3521

The next step is to establish several equations that can be solved simultaneously to determine the value of k. These equations are empirical in nature, as follows:

The y-intercept of the trendline on Figure 2.2 is equal to the value $(kL)^{1/3}$,

Where:

k = Reaction rate constant being sought
L = Ultimate (or initial) carbonaceous BOD concentration

In this case, $(kL)^{-1/3} = 0.21$
$kL = (0.21)^{-3} = 107.98$
$L = 107.98/k$

The second equation is the equation of the slope of the trendline on Figure 2.2. The vertical dimension of that slope at any point is equal to $k^{2/3}$. The horizontal dimension at the same location is equal to $6L^{1/3}$. Thus, the slope of the line at any point is equal to:

$$Slope = k^{2/3}/6L^{1/3}$$

By examination of the graph or the table earlier, it can be seen that the vertical component of the slope of the line between 2.7 and 3.7 days (a convenient horizontal distance of 1) is $0.2568 - 0.2372 = 0.0196$. Thus, the slope of the line = $0.0196/1 = 0.0196$.

Therefore:

$$k^{2/3}/6L^{1/3} = 0.0196$$
$$k^{2/3} = (6)(0.0196) L^{1/3} = 0.1176 L^{1/3}$$
$$k = (0.1176)^{3/2} (L)^{\frac{1}{2}}$$
$$k^2 = (0.1176)^3 L = 0.00163 L$$
$$k^2 = (0.00163)(107.98/k)$$
$$k^3 = 0.1760$$
$$k = (0.1760)^{1/3}$$
$$\underline{k = 0.561/day}$$
$$L = 107.98/0.561 = 192.5 \text{ mg/L}$$

Using the equation for a nonlag curve on these data would yield a value of $k = 0.2436$. Thus, it is important to know whether there is or is not a lag in the BOD data.

The Least Squares method involves fitting a curve through a series of data points such that the sum of the squares of the difference between the observed data and the values on the fitted curve are at a minimum. This can yield a variety of different curves for any data set and typically results in a differential equation to solve to yield the k-rate.

In the Fujimoto method a plot is made of BOD_{t+1} over BOD_t, both in mg/L. A line of slope 1.0 is then drawn from the origin of the graph; where the two lines intersect corresponds to the ultimate BOD of the waste (L) and the k-rate is determined from the BOD equations shown earlier as Equations 2.1 through 2.6.

The Respirometer method is a complex laboratory process that utilizes large volume electrolysis respirometers. Oxygen is maintained at a constant pressure over the sample during the procedure. Make-up oxygen is generated by an electrolysis reaction to replace oxygen used by the microorganisms. The BOD utilized is determined from the length of time oxygen is generated to make up for use and then correlating those data to the amount of oxygen produced by the electrolysis reaction.

2.13   NITROGENOUS BOD

After 5 days, the BOD exertion curves generally flatten out very quickly and very little additional oxidation is possible. The exertion of oxygen

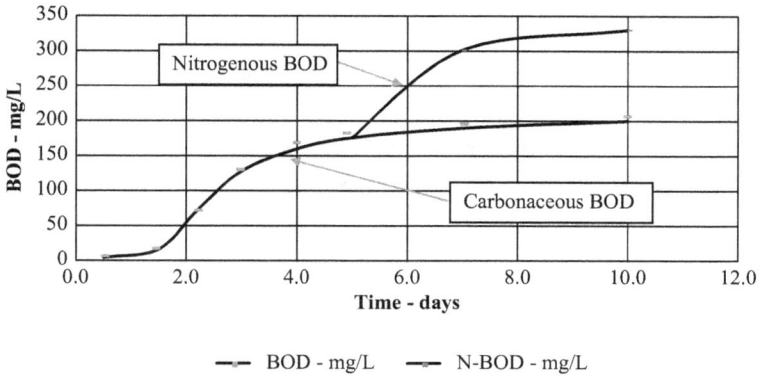

Figure 2.3. Carbonaceous BOD with nitrogenous BOD curve.

demand is not over at that time however. Generally, between about day 5 and day 8 a new oxygen demand curve, very similar to the original curve, will develop. This is a nitrogenous oxygen demand curve and is not related to the oxidation of organics. The development of the nitrogenous demand curve and the loss of appreciable gain in the organic oxygen demand curve are the principle reasons the BOD test is standardized at 5 days (see Figure 2.3). Although there are ways to suppress the nitrogenous BOD, it is generally assumed in practice that the BOD exerted within the first five days is essentially all carbonaceous BOD, often reported a cBOD, as seen in Figure 2.3.

2.14   TEMPERATURE EFFECTS ON k-RATE

Biological reactions are equally as dependent on temperature as chemical reactions. The k-rate is corrected for temperature using the following standard correction formula:

$$k_t = k_{20}\, \theta^{t-20} \tag{2.11}$$

where θ has an average value between 1.072 and 1.047 depending on the temperature at which the original BOD curve was developed. This correction factor assumes that the original k-rate was determined from a BOD curve that was developed using a 20°C incubation temperature. If that is not the case, the 20 in the power of theta must be adjusted to the actual temperature at which the BOD test was conducted and the value of theta must be adjusted accordingly. It is noted that the value of theta in sludge digestion processes has been reported to be significantly lower than that for the wastewater itself. Theta in the range 1.05 has been reported for sludge digestion. The actual value in all cases depends heavily on the concentration of the various constituents, the ratios between the various constituents, the pH of the matrix, and a variety of other, more difficult to define, parameters. The value 1.047 is typically used for wastewater work unless a different value is indicated by early results.

2.15   BIOLOGICAL GROWTH CURVE KINETICS

The growth of microorganisms in a substrate is a function of the substrate, the temperature, and the biomass concentration. Figure 2.4 shows a typical growth curve for wastewater organisms.

Without going into this diagram in detail, it is noted that the declining growth phase is the most efficient for secondary wastewater treatment due to the better settling characteristics of the biomass in this stage relative to the exponential growth phase, where instinct would expect the removal rate to be most efficient due to the more rapid uptake of the substrate. The location of the growth phase of the biomass along this curve is controlled by the recycle rate in the secondary stage of the wastewater treatment system and that rate controls what is called the food-to-microorganism ratio, or F/M ratio, in the reactor. The entire efficiency of the secondary treatment plant is controlled by maintaining the biological system within this declining growth portion of the growth curve.

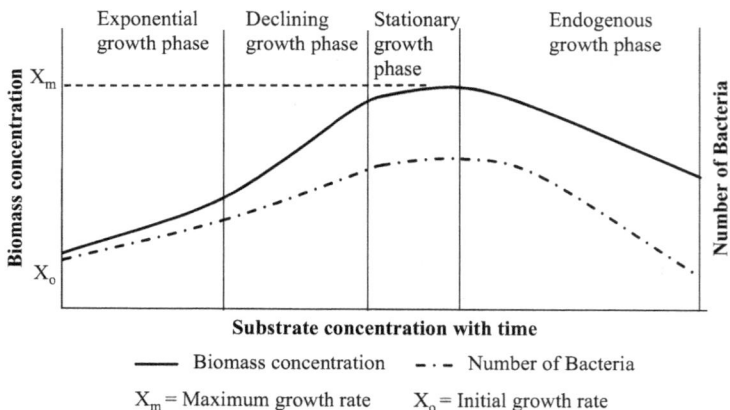

Figure 2.4. Typical growth curves for bacteria.

See Chapter 3 for further details on controlling the secondary treatment system.

2.16 DISSOLVED OXYGEN CONCEPTS, MEASUREMENT, AND RELEVANCE

Oxygen is the key ingredient in the degradation of organic matter in wastewater treatment plants. In order to be useful to the organisms, the oxygen has to be dissolved, although the oxygen contained in various compounds and components of the wastewater may also be available, that oxygen is far more difficult for the organisms to access than the dissolved form. The dissolved form of oxygen, or DO, is measured in terms of mg/L. There are lots of different, commercially available, DO meters that can measure the DO concentration, in real time, with good accuracy. Continuous DO monitors are routinely placed in key locations within wastewater treatment plants to help monitor the health of the treatment processes.

Table 2.2. Maximum DO concentration with temperature

°C	°F	DO mg/L	°C	°F	DO mg/L	°C	°F	DO mg/L
0	32	14.61	17	62.6	9.66	34	93.2	7.05
1	33.8	14.20	18	64.4	9.46	35	95.0	6.94
2	35.6	13.82	19	66.2	9.27	36	96.8	6.83
3	37.4	13.45	20	68.0	9.08	37	98.6	6.72
4	39.2	13.10	21	69.8	8.91	38	100.4	6.61
5	41.0	12.76	22	71.6	8.73	39	102.2	6.51
6	42.8	12.44	23	73.4	8.57	40	104.0	6.41
7	44.6	12.13	24	75.2	8.41	41	105.8	6.31
8	46.4	11.84	25	77.0	8.25	42	107.6	6.22
9	48.2	11.56	26	78.8	8.10	43	109.4	6.13
10	50.0	11.28	27	80.6	7.96	44	111.2	6.04
11	51.8	11.02	28	82.4	7.82	45	113.0	5.95
12	53.6	10.77	29	84.2	7.68	46	114.8	5.83
13	55.4	10.53	30	86.0	7.55	47	116.6	5.74
14	57.2	10.30	31	87.8	7.42	48	118.4	5.65
15	59.9	10.08	32	89.6	7.29	49	120.2	5.57
16	60.8	9.86	33	91.4	7.17	50	122.0	5.48

The amount of oxygen, and thus the resulting concentration of oxygen, that can be maintained in water is most directly a function of temperature, among other things. The maximum concentration of DO possible in pure water has been determined many times. Table 2.2 shows generally accepted values for the maximum concentration of oxygen in pure water at atmospheric pressure.

Various sources will show slight variations in the maximum DO concentration with temperature. That is often a function of the salinity of the water and the atmospheric temperature at the time the readings were made. Table 2.2 shows average values with the differences generally occurring in the second decimal place with a value variation of 1 or 2.

2.17 BIOLOGICAL NITRIFICATION AND DENITRIFICATION

Nitrogen is removed from wastewater through a two-step process involving the biological conversion of ammonia nitrogen to nitrate nitrogen and then denitrification through the biological conversion of the nitrate to nitrogen gas.

Most of the organic nitrogen that arrives at a wastewater treatment plant arrives in the form of ammonia created through the process of hydrolysis during the time the material travels in the sewer pipes. The largest and most common source of that organic nitrogen is the fecal matter and urea disposed into the sewers.

The conversion of ammonia to nitrate is an aerobic biological process. This process requires the oxidation of ammonium to nitrite and the secondary oxidation of the nitrite to nitrate. This first conversion is generally done by *Nitrosomona* bacteria, while the second is generally done by *Nitrobacter* bacteria. The two reactions occur simultaneously and proceed rapidly to the nitrate stage. Consequently, nitrite concentrations are generally low in wastewater entering the treatment plant.

Both of these bacteria are strict aerobes, meaning that they require free oxygen to effect the conversions. In addition, the nitrification process requires a long detention time, a low food-to-microorganism (F/M) ratio, and a long mean cell residence time. The process produces various acids such that an appropriate pH with adequate alkaline buffering is required for the process to reliably proceed. The optimum pH for these reactions is between 7.5 and 8.5, although nitrification has been reported at pH values of 6.5 to 7.0. About 7.1 mg/L of alkalinity, as $CaCO_3$, are required for each mg/L of ammonium nitrogen oxidized.

Temperature plays a part in these reactions but has not been shown to be a significant influence on the rate of reaction. Nitrification appears to reach a peak between 30°C and 35°C, but that it drops off rapidly after about 40°C. Nitrification appears to proceed slowly down to a temperature of about 10°C, but below 10°C it declines rapidly to zero. Denitrification occurs over a similar temperature range, with the reaction rate increasing as the temperature rises through that range.

The conversion of nitrate to nitrogen gas, the denitrification process, is accomplished by facultative, heterotrophic bacteria. The facultative bacteria use the oxygen attached to the nitrate (NO_3) as an energy source. This occurs when the DO concentration is so low as to all but shut down aerobic bacterial activity, creating what is called "anoxic" conditions. In aerobic conditions, the facultative bacteria use the DO rather than go through the trouble of dissociating the nitrates. In anoxic or anaerobic conditions, they will break down the nitrates and allow the nitrogen gas to escape from the liquid.

Denitrification occurs at a pH between 7.0 and 8.5. It also generates alkalinity as a by-product off-setting about half the alkalinity needed for the nitrification process. There also needs to be a sufficient source of available carbon for the denitrifying bacteria since there is none available from the nitrates. There are generally sufficient quantities of carbon available in the wastewater, but that may require some supplemental additions of carbon, such as methanol or acetic acid, in situations unusually high in ammonium or when denitrification occurs following secondary treatment. Secondary treatment tends to utilize the available organic carbon entering the plant with the wastewater and the carbon augmentation stage is required to provide enough carbon for the denitrifiers to work effectively.

Most nitrification/denitrification reactions are done in a two-stage reaction that allows for complete oxidation of the ammonia to nitrate in an aerobic reactor, followed by denitrification in an anoxic reactor. Recycling a certain percentage of the material from the anoxic reactor to the aerobic reactor allows for sufficient detention time to effect good results. The total detention time required is a function of the ammonium concentration, the temperature of the water, the availability of a suitable carbon source, and the alkalinity buffering capacity present.

BIBLIOGRAPHY

Anonymous. 2013. "Celcius-to-Fahrenheit-Table." http://www.metric-conversions.org/temperature/celsius-to-fahrenheit-table.htm (June 2, 2013).

Anonymous. 2013. "Solubility-of-Oxygen-in-Water." http://www.insiteig.com/pdfs/solubility-of-oxygen-in-water.pdf (June 2, 2013).

Flegal, T.A. 1976. "Temperature Effects on BOD Stoichiometry and Oxygen Uptake Rates." *Water Pollution Control Federation* 48, no. 12, pp. 2700–2707.

Hammer, M.J. 2004. *Water and Wastewater Technology.* 5th ed. Upper Saddle River, NJ: Prentice-Hall, Inc.

Hawley, G.G. 1981. *The Condensed Chemical Dictionary.* 10th ed. New York, NY: Van Nostrand Reinhold Company.

Metcalf & Eddy, Inc. 2003. *Wastewater Engineering: Treatment and Reuse.* 4th ed. New York, NY: McGraw-Hill Publishers.

Novak, J.T. 1974. "Temperature-Substrate Interactions in Biological Treatment." *Water Pollution Control Federation* 46, no. 8, pp.1984–1994.

Randall, C.W. 1975. "Temperature Effects on Aerobic Digestion Kinetics." *Journal of the Environmental Engineering Division of the American Society of Civil Engineers,* 101, no. 5 pp.795–811.

Randall, C.W. 1992. *Design and Retrofit of Wastewater Treatment Plants for Biological Nutrient Removal.* Lancaster, PA: Technomic Publishing Company, Inc.

Spector, M. 1998. "Cocurrent Biological Nitrification and Denitrification in Wastewater Treatment." *Water Environment Research* 70, no. 7, pp. 1242–1247.

Weaver, G. 2013. "Nitrification and Denitrification." http://www.cleanwaterops.com/wastewater-science/ (June 3, 2013).

CHAPTER 3

WASTEWATER TREATMENT PROCESSES

3.1   INTRODUCTION

Wastewater treatment is generally performed in a series of well-choreo-graphed stages. These stages are called

- preliminary treatment;
- primary treatment;
- secondary treatment;
- tertiary (or advanced) treatment.

In addition, all wastewater treatment plants generate waste solids, vari-ously referred to as "sludge," "biosolids," or similar terms, that need to be managed. They almost all also require disinfection of the effluent prior to discharge. Sludge management and disinfection are explained in detail later.

Preliminary treatment generally consists of screening to remove large floating objects, rags, and other things that could damage plant equip-ment; flow measurement devices; storage facilities to even out the flow to the plant; and grit removal to take out the large gravel, stones, and other, mostly inorganic, components that get into the system.

Primary treatment consists of a sedimentation basin in which rela-tively heavy objects settle out and buoyant materials, such as plastic, as well as fats, greases, and oil, float to the top. These are mostly organics at this stage, but there may be a few inorganics mixed in with them, as well.

The secondary portion of the treatment plant is a biological system of some kind, which is discussed in detail later, followed by a secondary sed-imentation basin and a recycle system. Wasting, or removal and disposal of excess solids developed in the secondary treatment system, is a key component of the process and the rate at which solids are recycled to the inlet to the secondary system controls the entire treatment process.

The tertiary system includes nutrient removal or other more esoteric treatment operations. It is not needed at all on some plants.

3.2   BASIC DESIGN PARAMETERS

The overall objective of wastewater treatment is to remove enough solid and dissolved organic material that the wastewater is safe for discharge to the intended receiving water or other media (that can be rivers, streams, ponds, lakes, oceans, or groundwater—or overland). It is generally *not* intended to provide potable or even swimmable or fishable water except in very rare cases.

3.2.1   NATURE OF WASTEWATER

To achieve the design objectives, it is first necessary to understand the nature of raw sewage and to understand the required discharge quality of the water when it leaves the wastewater treatment plant. Wastewater, for purposes of this discussion, includes everything that comes down the sewer pipe and enters the wastewater treatment plant. That can include: domestic sewage (from households and apartments), storm runoff, commercial and industrial wastes, infiltration and inflow water, and things that people deliberately flush down the sewer, including hazardous household wastes and other toxic substances they want to get rid of.

Sewage is by definition a combination of wastewater from all of those sources. It refers to a mixture of wastewater flows from residential properties, commercial properties, municipal properties, and industrial properties, plus infiltration and inflow. It also includes debris that gets into the pipes from a variety of sources, including rags, stones, rocks, boards, tree branches, occasional discarded fish, and small animals.

The volume of wastewater is estimated by various sources at about 120 gallons per capita per day (gpcd) from all sources combined. Wastewater flows from individual households, referred to as "domestic wastewater," vary widely and are dependent upon the number of residents, the location within the country, the weather at any given moment, the economic status of the community, and the number and types of commercial operations in the community, among other factors. A value of 60 to 80 gpcd for household flows is generally considered reasonable within the United States. Anticipated commercial flows are added to that value.

The organic loading (suspended solids and 5-day Biochemical Oxygen Demand [BOD_5]) is generally found to be about

0.24 lbs. of suspended solids per person per day; and
0.20 lbs. of BOD_5 per person per day.

These values assume water saving toilets and appliances in all households, and garbage grinders in the kitchen.

Table 5.1, in Chapter 5, is similar to the tables found in many sewer ordinances to provide guidance on sizing subsurface wastewater treatment systems. These are useful guides to sizing treatment plant components, as well, when there are no actual data available for this purpose.

Table 3.1 describes the approximate chemical makeup of "average" raw sewage, based on about 200 gpcd for low strength wastewater, about 120 gpcd for medium strength wastewater, and about 60 gpcd for high strength wastewater. These values are a compilation of data from several sources, not all of which are fully consistent with each other.

Industrial wastes can be highly variable and totally incompatible with municipal wastewater treatment plants. Most communities, therefore, require industrial pretreatment prior to discharge of industrial wastewater to the municipal sewer. This pretreatment is designed to convert the industrial waste to a form compatible with the municipal treatment. The characteristics of industrial waste are so variable that any kind of "average" numbers for any "normal" discharge parameters would be meaningless. It is necessary to evaluate each industrial discharge separately.

Infiltration and inflow (I/I) are two big contributors to wastewater flow, particularly in older systems. Infiltration is groundwater that leaks into the pipes from places like leaky sewer joints and breaks in the pipe. Inflow is surface water that leaks into the pipe through faulty manhole covers, combined sewer connections, and illegal cellar or drain connections. New sewer systems are generally designed on the basis of an average daily I/I of 100 gpcd, but are designed to carry up to 400 gpcd as they age.

The primary system removes about 35 percent of the incoming BOD_5, as a rule, and about 50 percent of the suspended solids. The biological portion of the secondary system removes an additional 50 percent of the initial BOD_5, an additional 35 to 40 percent of the suspended solids, and about 85 percent, or more, of the total dissolved solids. Conventional treatment, through the secondary stage, seldom removes significant quantities of nitrogen, phosphorous, or other nutrients. Tertiary treatment is necessary to achieve those reductions where required by discharge permits.

Table 3.1. Comparative composition of raw wastewater

Constituent	Low strength	Medium strength	High strength
5-day Biochemical Oxygen demand (BOD_5), mg/L	100–120	190–220	350–400
Chemical Oxygen Demand (COD), mg/L	240–260	430–500	800–1,000
Chlorides, mg/L	28–32	48–52	90–100
Fecal coliform per 100 mL	10^3–10^5	10^4–10^6	10^5–10^7
Fixed dissolved solids as a % of total dissolved solids	59–61%	59–61%	59–61%
Fixed suspended solids as a % of total suspended solids (TSS)	19–21%	19–21%	19–21%
Free ammonia nitrogen, as a % of total nitrogen	59–61%	61–63%	61–65%
Inorganic phosphorous as a % of total phosphorous	74–76%	71–72%	83–84%
Oil and grease, mg/L	48–52	90–100	100–150
Organic nitrogen as a % of total nitrogen	39–41%	37–39%	35–37%
Organic phosphorous as a % of total phosphorous	24–26%	28–29%	16–17%
Sulfates, mg/L	18–22	28–32	48–52
Total coliform per 100 mL	10^6–10^8	10^7–10^9	10^7–10^{10}
Total dissolved solids, mg/L	69–71	69–71	69–71
Total nitrogen, as N, mg/L	19–21	35–41	70–85
Total organic carbon (TOC), mg/L	75–85	140–160	260–290
Total phosphorous, as P, mg/L	3–5	7–8	12–15
Total settleable solids, mg/L	4–6	9–11	19–21
Total solids, mg/L	300–400	700–800	1,200–1,250
Total suspended solids (TSS), mg/L	100–120	210–220	350–400
Volatile dissolved solids as a % of Total Dissolved Solids (TDS)	39–41%	39–41%	39–41%
Volatile Organic Compounds (VOC), mg/L	50–75	100–400	500–750
Volatile suspended solids as a % of TSS	75–81%	75–81%	75–81%

3.2.2 EFFLUENT WATER QUALITY REQUIREMENTS

For wastewater treatment plants built within the United States, wastewater treatment plant effluent quality standards are set by the United States Environmental Protection Agency (EPA). Those regulations, found at 40 CFR 122 and 40 CFR 133, include limitations on the average concentration of suspended solids, BOD_5, pH, temperature, grease, and oil, and other constituents that depend upon specific industry discharges. Most states have now been delegated authority from the EPA to enforce the federal regulations as part of their enforcement of state and local regulations. These requirements are all built into a National Pollutant Discharge Elimination System (NPDES) permit issued to each treatment plant. All NPDES permits include all of the federal requirements and also include a significant number of local regulations that depend upon the water quality in the receiving water and how the various states regulate that water quality.

For treatment facilities in other parts of the world, water-quality based standards are most often used, where such standards have been developed by the appropriate regulatory authorities. Where governmental standards have not been developed or adopted, general water-quality based standards are still appropriate as a means of establishing effluent discharge concentrations. In those cases, health-based concentrations may also be applicable using World Health Organization or other appropriate health standards to define effluent quality parameters.

Table 3.2 is adapted from the Code of Federal Regulations (CFR), specifically the United States Environmental Protection Agency (EPA) regulations at 40 CFR 133 and provides a list of federally regulated discharge limits, as of December 2013. Federal regulations are not static and permit limits may change over time. Designers are advised to check the current regulations to verify current requirements when designing new facilities or upgrading existing facilities.

It is noted that chlorine residuals are not a standard discharge limit established by EPA. The numbers shown in Table 3.2 are from the Massachusetts Water Resources authority discharge permit for the Deer Island wastewater treatment plant, which discharges to Boston Harbor. Some permits require only seasonal disinfection of discharges and others may have strict limits on chlorine residuals due to potential impacts on wildlife. Those facilities are often fitted with dechlorination systems to reduce the chlorine residuals or they are retrofitted with ultraviolet light or ozone systems for disinfection, which leave no residual.

The EPA regulations at 40 CFR 133.105 also provide for variations in the preceding numbers based on a determination that a facility is eligible

Table 3.2. EPA discharge limits for wastewater treatment plants (40 CFR 133.102)

	Total and carbonaceous (cBOD$_5$) BOD$_5$	Suspended solids	Oil and grease*	pH	Fecal coliform**	Total chlorine residual***
Monthly average	30 mg/L BOD$_5$ or 25 mg/L cBOD$_5$ or 85% removal	30 mg/L or 85% removal	10 mg/L	6.0–9.0	200–14,000 colony forming units per 100 mL	456 ug/L
Weekly average	45 mg/L BOD$_5$ or 40 mg/L cBOD$_5$	45 mg/L		6.0–9.0	200–14,000 colony forming units per 100 mL	
Daily maximum				6.0–9.0	200–14,000 colony forming units per 100 mL	631 ug/L

*Oil and grease is not a secondary treatment standard per 40 CFR 133. This value is a typical state imposed value.
**Number depends on permit. 200 is "normal"; the 14,000 comes from the Massachusetts Water Resources Authority permit limits for discharge to Massachusetts Bay.
***Total chlorine residual values are not required by all NPDES permits. Limits shown are from the Massachusetts Water Resources Authority permit limits for discharge to Massachusetts Bay.

for treatment equivalent to secondary treatment. Such facilities are defined at 40 CFR 133.101(g) as:

(g) Facilities eligible for treatment equivalent to secondary treatment. Treatment works shall be eligible for consideration for effluent limitations described for treatment equivalent to secondary treatment (§ 133.105), if:

1. The BOD5 and SS effluent concentrations consistently achievable through proper operation and maintenance (§ 133.101(f)) of the treatment works exceed the minimum level of the effluent quality set forth in §§ 133.102(a) and 133.102(b).
2. A trickling filter or waste stabilization pond is used as the principal process, and
3. The treatment works provide significant biological treatment of municipal wastewater.

Facilities eligible for these revised standards must meet the discharge limits shown in Table 3.3.

Various special considerations for certain industries and certain types of uncommon treatment facilities are also provided at 40 CFR 133.103.

Tables 3.4, 3.6, 3.8, 3.9, and 3.10 provide design criteria for various unit operations within a wastewater treatment facility. For the most part, these design parameters and units are fairly obvious, with a little forethought. Some parameters show multiple values based on different sources. However, most design standards for wastewater treatment plants were developed by the Great Lakes Upper Mississippi River Board of State and Provincial Public Health and Environmental Managers (GLUMRB) from empirical data developed over many years. The work of that organization has been found to be exceptionally good and has evolved into the standards of operation for most regulatory authorities, except where local

Table 3.3. EPA discharge limits for wastewater treatment plants eligible for treatment equivalent to secondary treatment (40 CFR 133.105)

	BOD_5	Suspended solids	pH
30-day average	45 mg/L	45 mg/L	6.0–9.0
7-day average	65 mg/L	65 mg/L	6.0–9.0
30-day average % removal	65%	65%	

experimentation has shown that local conditions are sufficiently different from those in the GLUMRB area, that local variations are justified.

3.3   PRELIMINARY TREATMENT UNITS

Preliminary treatment consists of screening and shredding, grit removal, flow measurement, flow equalization (where needed), raw sewage pumping, and septage receiving stations. Each of these operations is discussed in detail in the following text and the design criteria are provided in Table 3.4.

3.3.1   SCREENING AND SHREDDING

Pretreatment screening is used to protect downstream equipment from the problems associated with rags and other solid objects in wastewater. In several modern plants, especially those that have floating or neutrally buoyant media with screens included in the secondary and tertiary system (to keep the media out of the receiving water), fine screens are required for preliminary treatment to take out even small solids that would "blind" the downstream process. Fine screening ahead of some processes sometimes require perforated plate-type units, which are not really screens at all, but which serve a similar function.

Screens for sewage come in a variety of types and opening sizes. Coarse screens remove gross debris, such as leaves, sticks, boards, and similar objects, which find their way into the system. These screens typically have openings of about ½ to 1 inch in width and are made of very sturdy stock, typically bars of steel on edge to the flow of water. They are typically ½ inch across and up to 3 inches deep, set ½ to 1 inch apart. Coarse screens are generally cleaned with mechanical rakes. These devices have bars that extend between the bars of the screen that move upward to drag debris over the top of the screen and deposit it into a moveable hopper. The rake then swings out and drops back down to the bottom of the screen to begin a new, slow rise to the top.

Fine screenings are often difficult to handle and manage because they tend to be putrescible, they tend to be already decaying before they arrive at the treatment facility, and they are grossly odorous. Fine screens are usually comprised of moving belts with openings between 1/8 and ¼ inches wide. They are fitted with a doctor blade to scrape debris off of the screen and drop it into a moveable hopper. They also generally contain a backwash system to wash fine debris that tends to clog the screens back into the liquid stream going to the primary portion of the treatment plant.

Table 3.4. Typical design criteria for various preliminary wastewater treatment plant unit operations

Unit operation	Approach velocity	Flow rate	Flow through velocity	Location	Basin geometry	Mechanical mixing requirements	Aeration requirements	Detention time
Screening and shredding	1.5–3.0 ft/sec (0.4–0.9 m/sec)			First unit operation to provide protection for downstream units	Slope from vertical at 30°–45° for manually cleaned screens, <30° for mechanically cleaned screens			
Grit removal (aerated grit chamber)			0.5–1.5 ft/sec across the bottom (0.15–0.45 m/sec)	Second unit operation to minimize wear on downstream equipment	Bottom slope of 45°; depth of 6–15 feet (1.8–4.5 m);width to liquid depth ratio of 2:1 up to 4:1; length to width ratio of 2:1–5:1; length 25–90 ft (7.6–8.3 m)		0.02–0.15 ft³/sec per ft of basin length (0.002–0.012 m³/sec per m of basin length)	2–5 minutes at peak hourly flow (typically designed for 3 minutes)

(Continued)

Table 3.4. (*Continued*)

Unit operation	Approach velocity	Flow rate	Flow through velocity	Location	Basin geometry	Mechanical mixing requirements	Aeration requirements	Detention time
Flow measurement (Parshall flume)				Following screens and grit removal to indicate the flow into the primary units most accurately	See Figure 3.1			
Flow equalization		1.1–1.25 times the average diurnal flow variation		Downstream of screens and grit removal, prior to flow equalization; offline basins are used to attenuate storm flows while inline basins attenuate dry weather flow	Continuous flow, stirred tank, 4.5–6.0 ft deep (1.5–2.0 m). Aerated to maintain aerobic conditions and to minimize sedimentation inside the basin	0.015–0.030 kW/ft^3 (0.004–0.008 kW/m^3) of basin storage capacity at approximately 200 mg/L SS	0.6–0.9 ft^3 of air per cubic foot of volume (m^3 of air/m^3 of volume)	

Raw sewage pumping	After flow equalization to regulate and stabilize flow into treatment units	
Septage load to primary	Offline with controlled feed to flow equalization basin	Aerated to begin conversion to aerobic conditions. That will minimize negative impacts on the primary treatment units

In both cases, it is common to discharge the screenings from the moveable hoppers into a washing basin where the organic material is washed off the inorganic material and the organics are returned to the plant inlet. This is done to reduce the organic load on the landfill site where the inorganics are disposed and it reduces the putrification potential for the screenings. This is a violation of the basic rule of wastewater treatment, which says "once out, keep it out." It is a useful point to vary that rule, however, since the rest of the treatment plant is designed to handle the organic fraction, the fraction returned is already fine enough to be effectively treated in the biological treatment portions of the plant, and the negative environmental effects from the landfill disposal of organic fractions could be significant. Large organics that cannot be effectively treated in the biological system are removed from the washing process and burned or landfilled with the inorganic fractions.

Shredders, grinders, or comminuters (all very similar pieces of equipment) crush, chip, or otherwise break up the larger incoming objects and make them small enough for the primary and secondary portions of the treatment plant to handle efficiently. It is important to place these devices after the screens to avoid large stones or other objects that cannot be effectively chipped from entering these devices. There is a huge amount of torque used to crush the incoming objects with most of this equipment and sudden stoppages due to uncrushable debris can cause serious damage to the equipment and injury to the operators.

This equipment does not actually remove anything, however, and so they are not used as often as screens. Where they are used, a by-pass channel is essential in case there is a malfunction or breakdown of the shredder or comminuter. Screens are used in the by-pass channel in the event the shredder or comminuter needs to be taken off line for repairs. Some plants screen out the larger objects, then put them through a grinder or comminuter and return them to the waste stream prior to primary sedimentation. This author does not recommend that approach; but prefers to utilize the "once out, keep it out" approach to these kinds of objects. Most are difficult to treat, even when shredded and add nothing advantageous to the downstream treatment units.

3.3.2 GRIT REMOVAL

Grit removal is intended to remove the heavy inorganics (and some organic components) from the waste stream so that they will not be able to damage the downstream facilities including wearing out raw wastewater pumps, overloading and breaking primary sludge flights (particularly in plants treating combined wastewater and stormwater), wearing out sludge pumps and changing the sludge characteristics, and then taking up space in digesters, which becomes very difficult to clean out. Grit is a major concern at wastewater treatment plants, and grit removal systems are often poorly designed.

Grit generally consists of heavy objects with a specific gravity that prevents it from being carried by the flow of the water when that flow is reduced to values below about 2 ft/sec or so. This includes minor volumes of sand and gravel that have not settled out in the collection system, plus coffee grinds, egg shells, and other coarse, heavy materials that tend to settle quickly and easily when the stream flow is rapidly reduced. Grit chambers are designed for the peak hourly flow rate and a detention time of about 1 minute at that flow rate. The tanks are generally shallow square units with forced aeration added to keep the organics in suspension while the heavy inorganics settle out. This aeration also tends to agitate the material, allowing the particles to grind against each other, further cleaning the settleable materials and keeping as much of the organic fraction as possible in the waste stream. This makes the materials removed less putrescible and easier to handle. The grit is generally landfilled.

The mechanism for grit removal can be either a basin in which the flow rate is suddenly reduced to less than 2 ft/sec; a similar basin that is aerated to create a swirling effect that can help to separate the grit from treatable organic material that may be attached to it, or vortex-style degritters. In all cases, the objective is to slow down the flow rate sufficiently such that the heavier objects settle out while the lighter, organic fractions carry over into the rest of the treatment plant.

3.3.3 Flow Measurement

Flow into or out of a wastewater treatment plant is generally in an open pipe or channel. For measuring wastewater flow, a Parshall flume, as shown in Figure 3.1, has historically been utilized. With this device, the flow is measured as a function of the head differential through a controlled restriction in the flume.

The equation for a flume is shown as Equation 3.1.

$$Q = 4Bh^{1.522B^{0.026}}$$

(3.1)

Where:

Q = flow in cubic feet per second (cfs)
B = throat width in feet
h = upper head in feet

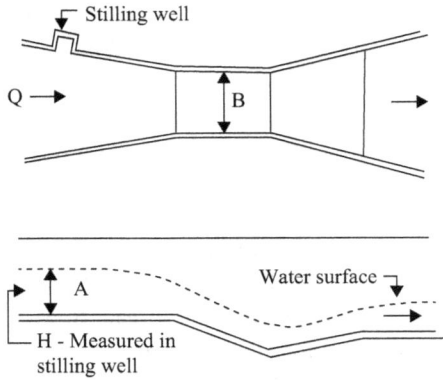

Figure 3.1. Cross-sectional view of a typical Parshall flume.

This formula is not generally convertible to metric units without changing the constants. It is noted that several texts show the power of B in the power of h as 0.062. This is an incorrect typographical error that showed up in an early text and was carried over by reference to several later texts. The correct value is 0.026, as shown earlier.

Example Problems 3.1 and 3.2 demonstrate the use of this equation.

Example Problem 3.1

Calculate the flow rate through a Parshall flume with a throat width of 2.5 feet and a free-flow head of 3.0 feet

Solution

$Q = 4 \ (2.5 \ \text{ft}) \ (3 \ \text{ft})^{1.522(2.5) \ ^\wedge 0.026}$
$Q = 4 \ (2.5) \ (4.17)$
$Q = 55.4$ cubic feet per second (cfs)
$Q = 55.4 \ \text{cfs} * 86,400 \ \text{sec/day} = 4,786,560 \ \text{cf/day}$
$Q = 4,786,560 \ \text{cf/day} * 7.48 \ \text{gal/cf} = \underline{35,803,469}$ gal/day.

Example Problem 3.2

A wastewater flow has the following characteristics. Assume that the inlet open channel is 2.5 feet wide and the transition zone to the flume is 1.0 foot long. The maximum inlet velocity is 3 feet per second (ft/sec).

The flow rate is not constant, but varies constantly from a low flow rate of about 3.5 million gallons per day (mgd) at 3:00 am to a peak flow rate of about 6.0 mgd at 10:00 am. The flow then declines to about 5.0 mgd around 3:00 pm and back up to 6.0 mgd by 9:00 pm. The flow then drops off steadily to 3.5 mgd at about 3:00 am.

What is the value of "h" at the minimum flow rate and what is the flow velocity through the flume at that flow rate?

Solution

$$3.5 \text{ mgd} = (3.5 \times 10^6 \text{ gal/d}) \times (1 \text{ d/86400 sec}) \times (1 \text{ cf/7.48 gal})$$
$$Q = 5.42 \text{ cf/sec}$$

By Equation 3.1,

$$Q = 4Bh^{1.522B^{0.026}}$$
$$5.42 \text{ cf/sec} = (4 \text{ ft})(2.5 \text{ ft})(h)^{1.522(2.5)^{0.026}}$$
$$(h)^{1.559} = 5.42/12 = 0.452 \text{ ft}$$
$$1.559 \log h = \log 0.452 = -0.345$$
$$\text{Log } h = -0.345/1.559 = -0.221$$
$$h = 1.66 \text{ ft}$$
$$Q = VA$$

Where V = velocity of flow and A = cross-sectional area of the throat

$$(5.42 \text{ cf/sec}) = V(2.5 \text{ ft})(1.66 \text{ ft})$$
$$V = 5.42/4.15 = \underline{1.3 \text{ ft/sec}}$$

In more recent facilities, magnetic flow meters have been used on closed pipes, rather than an open channel flow device. Often a wet well is used at the influent to the plant and the wastewater is pumped to the top of the primary system so that the rest of the plant can operate with gravity flow between units, as much as possible. In those cases, the flow from the pumps may be measured as the treatment plant flow, or magnetic meters can be used on the pumped flow for the verification of flows.

3.3.4 FLOW EQUALIZATION

Wastewater flows into a treatment plant are traditionally highly variable over the course of a day. That is not unexpected. The times of the day that people are generating wastewater during the week are also highly variable. In addition, the times of day that flows are generated on weekends are different from those during normal week days. An observation of water demands in any community will indicate a large variation in the diurnal demand rates for potable water. Since domestic sewage is generated principally from the use of potable water, it should come as no surprise that wastewater flows are also highly variable diurnally. There is a marked difference, however, between the diurnal flow patterns of water demand and the parallel patterns of wastewater flow. That is because the wastewater has to flow relatively long distances in sewers before arriving at the treatment plant. In fact, in larger urban areas where huge treatment facilities serve very large population centers, sewage may remain in a sewer pipe for up to two or three days before actually arriving at the treatment facility. Consequently, the magnitude of the sewage flow variation is often mitigated by the travel time in the pipes. That is both a blessing and a curse; the flow is more constant, but the sewage will already be turning septic and require more severe immediate treatment to control odors and to minimize deleterious impacts on the treatment process.

Wastewater treatment plants generally operate most effectively when the flow into each of the unit operations is essentially constant and within about 75 to 100 percent of the design flow. If the influent to the facility varies significantly, it is difficult to maintain a constant flow through the treatment plant within a reasonable fluctuation rate throughout the day. Flow equalization can assist with this effort, but flow equalization costs money, takes up space, and can be odorous if not designed properly.

There are two fundamental ways to control flow into a treatment plant. The first, most common method, is to install a series of pumps in a large wet well that serves as a receiving station for all flows into the plant. These pumps are then set to operate at a constant, predetermined, flow rate to pump sewage into the primary treatment portion of the plant. If the flow variation into the facility is not so great as to require a huge wet well, which could be difficult to maintain, this concept can be effective and can reduce construction costs. The second method is to provide a separate wet well into which all flows initially are delivered. This basin will require aeration to maintain constant agitation of the wastewater, which will prevent settling of solids in the basin, and a separate, small wet well for the pumps that will deliver the wastewater to the primary system. The connection between the two wet wells requires a closable door or hatch so that in the event of trouble in the pump well, the flow can be stopped at the outlet from the flow equalization basin and that basin can accumulate flows for some predesigned period of time, typically up to 12 to 24 hours, while the problem with the pumping station is repaired. This concept also minimizes the use of by-pass channels that discharge untreated wastewater to a receiving stream in the event of major pump failure.

3.3.5 RAW SEWAGE PUMPING

Wastewater typically enters a treatment plant through an underground sewer. Occasionally, there is a force main that feeds the plant, but those usually discharge ahead of the open channel inlet so that all the flow can be measured through the same Parshall flume. It is convenient, however, to minimize the number of times flows are handled or pumped, so it is most common to pump the raw wastewater to a primary tank sufficient to provide adequate elevation head pressure to allow the wastewater to flow by gravity through all of the other unit operations and into the outfall pipe. There are significant head losses through all of the unit operations, pipes and valves, bends, and other features of the system. The total head required at the front end needs to be carefully considered for proper plant operation

The design of a pump is beyond the scope of this book. There are dozens of different kinds of pumps and each has a different capability and function depending on the size of the motor driving it and the size of the discharge and inlet pipes associated with it. The main factors in pump selection are: (*a*) head requirements (lift), (*b*) flow, (*c*) range of flow conditions expected, and (*d*) type of solids that will be found in the flow. The engineer needs to decide on the type of pump that best meets these conditions and to then work with pump manufacturers to get into the specific design. It is generally best to work directly with one or two pump manufacturers to ascertain the specific pump to recommend for any specific application.

3.3.6 SEPTAGE RECEIVING STATIONS

In rural areas, it is common to treat wastewater through subsurface disposal fields—septic systems (see Chapter 5). These systems consist of a septic tank in which most of the solids settle or float, a distribution box, and a series of laterals through which settled wastewater is distributed to the subsurface soils for treatment. The material that settles and floats in the septic tank must be pumped out periodically to minimize the opportunity for solids carry-over to the distribution field. Solids entering the distribution field will rapidly clog the distribution pipes and the soil surrounding them, causing complete failure of the disposal field.

When septic tanks are pumped, the material removed needs to be properly disposed. It is highly anaerobic, extremely odiferous, very high in BOD_5, high in suspended solids, and typically has a pH much higher than the normal incoming wastewater. It is also toxic to humans and needs to be carefully handled to avoid exposure to diseases and viruses. It is common to find separate, dedicated, grit removal and screening facilities for septage receiving stations. These facilities also typically incorporate storage facilities to allow excess septage to be blended slowly into the main wastewater flow stream so as not to disrupt the treatment processes with shock loadings of highly concentrated waste.

There are two considerations in the design of a septage receiving station. The first is a reasonable estimate of the amount of septage to be generated within the collection area of the facility and the second is the total volume of septage the facility can effectively handle.

Table 3.5 provides typical characteristics of domestic septage. Commercial septage is so highly varied and industry dependent as to defy rational estimates of average values. All septage should be tested for comparison

Table 3.5. Typical characteristics of domestic septage

Parameter	Reported range	Reported average	Suggested average value
pH	6.9–8.1	7.2–8.1	7.8
Total BOD, mg/L	165–78,600	303–10,000	3,600
COD, mg/L	181–703,000	668–28,200	10,000
Total Kjeldahl Nitrogen (TKN), mg/L	9–1,060	66–1,060	180
Ammonia nitrogen, mg/L	3–155	91–1,500	580
Phosphate phosphorous (PO_4-P), mg/L	5.4 to >39	13–39	25
Sulfates, mg/L	33–738	128–132	130
Hydrogen sulfide, mg/L	52 to >95	67–95	80
Total solids, mg/L	328–130,475	3,095–30,450	16,600
Total suspended solids (TSS), mg/L	76–93,378	3,068–14,600	9,400
Volatile suspended solids, mg/L	212–71,402	2,706–10,366	6,050
Oil and grease, mg/L	208–23,368	22,000–23,000	22,500

to the suggested and average values provided, however, since the ranges of values are extreme.

Septage generation rates have been reported in the range of 43 to 54 ft³ (322 to 404 gallons) per capita served per year (1.2 to 1.5 m³ or 1,200 to 1,500 L per capita served per year). General practice is to convert the septage strength to population equivalents (PE) based on an average of the equivalence between the septage values of 120 g/PE (COD), 60 g/PE (BOD), 12 g/PE (TKN), 70 g/PE (TSS), and 3 g/PE (PO_4-P). If there is capacity in the facility for that many extra people, the septage flow is metered proportionally to the incoming domestic flow from a separate septage flow equalization basin provided for that purpose. The metering pump can be electronically tied to the Parshall flume flow meter or manually adjusted on an hourly basis. Pretreatment of the septage to significantly reduce the BOD, COD, TKN, and PO_4-P to be compatible with the incoming concentrations is also common. The issue then is total loading, not population equivalents.

3.4   PRIMARY TREATMENT UNITS

3.4.1   PRIMARY SEDIMENTATION BASINS

The primary portion of a wastewater treatment plant is essentially a sedimentation basin—the cleverly named "primary sedimentation basin." Sedimentation basins can be circular or rectangular, but are designed to slow the flow of the wastewater sufficiently to allow the vertical settling velocity of the smallest particle for which 100 percent capture is desired to exceed the upward velocity of the water through which it is settling. The vertical velocity of the water equals the settling velocity of the target particle when they are both equal to the surface overflow rate (SOR) calculated as:

$$V_o = Q/A \qquad (3.2)$$

Where:

V_o = SOR in gallons per day per square foot of surface area (gpd /sf) or cubic meters per day per square meter (m/d)

Q = average daily flow in gallons per day (gpd) or cubic meters per day (m³/d)

A = surface area of the reactor in square feet (sf) or square meters (m²)

Detention time is not one of the standard design parameters for primary clarifiers. Nevertheless, normal detention times are typically in the order of 1.5 to 2.5 hours with approximately 2.0 hours being the most common.

Detention time is calculated by the standard equation of:

$$t_r = (24) (V/Q) \qquad (3.3)$$

Where:

t_r = detention time in hours
V = tank volume in 10^6 gals (m^3)
Q = average daily flow in millions of gallons per day (mgd) or (m^3/d)
24 = conversion factor to hours (hr/day)

Detention time can also be calculated in terms of hours using the water depth and overflow rate as follows:

$$t_r = (180 * H)/V_o \qquad (3.4)$$

Where:

t_r = detention time in hours
H = depth of water in the basin in feet
Vo = surface overflow rate in gallons per day per square foot
180 = a conversion factor equal to 7.48 gal/cf * 24 hr/day

Two things occur in sedimentation basins that need to be addressed. The first is the floating of light material to the surface of the basin and the second is settling of heavier materials to the bottom of the basin. To remove these materials, which tend to be organic and putrescible in nature, a rake is used. In a circular basin, which is most common for primary settling basins, the rake slowly rotates about the center of the basin skimming the floating materials onto a scum collection ramp from which the material slides into a trough and is removed by low volumes of water to a dewatering device. This rake also has a bottom bar attached to it that rides along the bottom of the basin scraping accumulated sludge into a hopper at the center of the basin from where it is periodically removed by opening the discharge valve and letting water pressure from above force the sludge into a sludge dewatering system. Typically the scum and the sludge are then combined and they are dewatered and treated further together.

With a rectangular basin, the rake is a continuous loop that skims the surface of the basin in one direction, then, after dumping the scum over the scum ramp, drops down to the bottom of the basin and scrapes the bottom sludge into a hopper at the opposite end of the basin before returning to the top for another cycle through the basin. A series of rakes on moving chains provides for continuous scum and sludge removal with a very slow movement of the rakes that will prevent re-entrainment of the collected solids on either pass.

The design ratios for width to length for rectangular basins is typically 3:1 to 5:1 with widths ranging from 10 to 20 feet and depths of 7 to 8 feet. If those dimensions will not provide an adequate SOR, the flow is divided into two or more trains to reduce the loading to each section such that they do meet the SOR criteria. SOR and weir loading are the primary design criteria for these basins. Retention time is not a standard criterion.

Weir loading is defined as the total flow divided by the total length of the overflow weirs. GLUMRB and the U.S. EPA both set standards for the design of these basins. Table 3.5 provides the current standards established by GLUMRB, the U.S. EPA, and the Water Environment Federation for the design of primary sedimentation basins.

Flow through velocity is a key part of the design of a clarifier to minimize scour of settled solids back into the effluent. Management of the design parameters shown in Table 3.5 will control the flow through velocity adequately, in most cases. The end result should be a flow through velocity in the order of 0.06 to 0.08 ft/sec (0.02 to 0.025 m/sec) to control scour.

Sidewall depth is defined by GLUMRB and EPA with the GLUMRB standard being the most conservative. Based on the set SOR and sidewall depths, the detention time can be calculated using Equation 3.5:

$$t = (180 * h)/(V_o) \qquad (3.5)$$

Where:

t = detention time in hours
h = sidewall depth in feet
V_o = SOR in gpd/sf
180 = conversion factor to fix the units.

Note that this equation does not work with metric units unless the conversion factor is first adjusted.

Example Problem 3.3 demonstrates the application of this equation.

Example Problem 3.3

Given a circular clarifier that is 96 feet in diameter with a sidewall water depth of 8.5 feet and a single effluent weir located along the periphery of the tank, determine the overflow rate, the detention time, and the weir loading for an average flow of 11 mgd.

Solution

First, calculate the surface area and volume of the clarifier.

$$\text{Surface area} = 2\pi r^2$$
$$= (2)\,(\pi)\,(96\ \text{ft}/2)^2$$
$$= 14{,}476\ \text{sf}$$
$$\text{Volume} = (14{,}476\ \text{sf})\,(8.5\ \text{ft})$$
$$= 123{,}050\ \text{cf}$$

Volume in gallons $= (123{,}000\ \text{cf})\,(7.48\ \text{gal/cf})\,(1\ \text{million gallons}/10^6\ \text{gallons})$
$$= 0.920\ \text{million gallons}$$

Surface Overflow Rate comes from Equation 3.2

$$V_o = Q/A$$
$$= (11{,}000{,}000\ \text{gallons/day})/(14{,}476\ \text{sf})$$
$$= \underline{760\ \text{gpd/sf}}$$

Detention time comes from Equation 3.3

$$t_r = 24\ V/Q$$
$$= (24)\,(920{,}000\ \text{gal})/(11{,}000{,}000\ \text{gal})$$
$$= \underline{2.0\ \text{hours}}$$

Detention time can also be calculated using Equation 3.4

$$t_r = [(180)\,(H)]/V_o$$
$$= [(180)\,(8.5)]/760$$
$$= \underline{2.0\ \text{hours}}$$

Weir loading comes from the circumference of the basin and the flow rate.

$$\text{Circumference} = 2\pi d$$
$$= (2)\,(\pi)\,(96\ \text{ft})$$
$$= \underline{603.2\ \text{ft}}$$
$$\text{Weir loading} = 11{,}000{,}000\ \text{gpd}/603.2\ \text{ft}$$
$$= \underline{18{,}236\ \text{gpd/ft}}$$

(a)

(b)

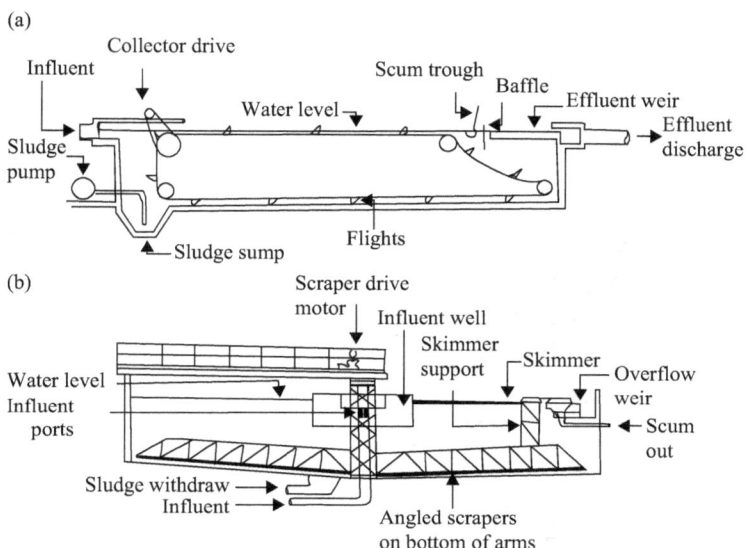

Figure 3.2. Schematic Views of a Rectangular (a) and a Circular (b) Clarifier.

Figure 3.2 shows schematic views of circular and rectangular primary clarifiers. Design criteria for both styles of primary clarifier are described in Table 3.6.

A more detailed discussion of the theory of sedimentation is found in Chapter 6. Primary sedimentation is principally discrete particle sedimentation; intermediary, secondary, and tertiary sedimentation tends to be flocculant settling.

It is noted that more attention has been paid recently to the solids loading rates for primary clarifier design. There is insufficient clarity on how those data are used to include in this edition. It is also noted, however, that enhanced solids removal has been demonstrated in primary clarifiers with chemical addition. Enhanced phosphorous removal has also been demonstrated with chemical addition to the primary clarifier. Both operations can also reduce the organic loading to the rest of the treatment plant and should be considered with care. Disposal of primary sludge with high organic content can be a problematic undertaking due to odors, putrification, and handling issues.

3.4.2 PRIMARY SLUDGE MANAGEMENT

Primary clarifiers are expected to remove from about 20 percent of the incoming BOD_5 when the incoming wastewater contains a significant

Table 3.6. Typical design parameters for primary clarifiers

Source	Overflow rate in gallons per square foot per day (gpd/sf)		Overflow rate in cubic meters per square meter per day (m/d)		Side wall water depth in feet (ft)	Side wall water depth in meters (m)	Weir loading in gallons per linear foot per day (gpd/ft)	Weir loading in cubic meters per linear meter per day (m³/d/m)
	Average monthly flow	Peak hourly flow	Average monthly flow	Peak hourly flow				
EPA	800–1,200	2,000–3,000	32.6–48.9	81.5–122.3	10–13	3.1–4.0	10,000–40,000	124.3–497.2
GLUMRB	1,000	1,500–2,000	40.8	61.1–81.5	7	2.1	10,000 for flows < 1 mgd; 20,000 for flows > 1 mgd	124.3
GLUMRB with secondary sludge recycle to primary clarifier		<1,200		<50				
EPA with secondary sludge recycle to primary clarifier	600–800	1,200–1,500	24.5–32.6	48.9–61.1	13–16	4.0–4.9		
Other sources[1] with secondary sludge recycling		1,000–1,700		40–70			10,000–15,000	120–190

[1]Metcalf & Eddy (2003) and WEF (1998)

Table 3.7. Common characteristics of untreated, settled, primary sludge

Moisture content, %	92–99 (typical 97)
Dry solids content, %	1–8 (typical 3)
% Organic matter in dry solids, %	60–80 (typical 70)
Nitrogen content of dry solids, %	1.5–4 (typical 2.5)
Volatile solids content of dry solids, %	60–85 (typical 75)
Phosphorous content of dry solids, %	0.8–2.8 as P_2O_5 (typical 1.6)
Potassium content of dry solids, %	0–1 as K_2O (typical 0.4)
pH	5–8 (typical 6)
Alkalinity, mg/L as $CaCO_3$	500–1,500 (typical 600)
Grease and fats content of dry solids, %	5–8 (typical 6)

Source: Metcalf & Eddy (2014) and Davis (2011)

concentration of soluble BOD_5 up to values 30 to 40 percent for "normal" domestic wastewater. Some industrial wastewater removal rates have been seen at 50 to 60 percent BOD_5 removal, but that is not common. Removal rates and resulting sludge densities are also a function of the hydraulic overflow rate of the clarifier. Higher overflow rates typically yield lower BOD_5 removal rates and lower sludge density values. See Table 3.7 for common primary sludge characteristics.

Primary sludge can be withdrawn on a continuous basin and sent directly to a sludge thickener for incorporation in the total sludge management system or they can be recycled to the inlet of the clarifier to accelerate the conversion of incoming organic solids to soluble BOD_5, which is required for the removal of the BOD_5 in the secondary system. A significant portion is always removed to a thickener at this point because the objective of sludge removal is to take the BOD_5 out of the flow stream. Recycle puts it right back in again. Therefore, the removal of the majority of the sludge to a thickener is necessary to reduce the overall loading to the secondary portion of the treatment plant. Sludge thickeners are discussed in Section 3.8.

3.5 SECONDARY TREATMENT

The secondary portion of a wastewater treatment plant consists primarily of a biological treatment unit of some kind, usually with a recycle component, followed by a secondary, or final, clarifier. The biological system has historically been one of two kinds—fixed film (such as Rotating Bio-logical Contactors [RBC] or trickling filters) or suspended growth (such as conventional activated sludge or a variation thereof). Some newer treatment systems combine both suspended and fixed film within a single tank, to increase the biomass within a limited footprint. This concept is actually becoming fairly common in newer facilities. Good, generic, design data are difficult to find and these concepts are not yet as well documented as the more historic concepts.

In any case, the growth of the biomass—mostly single celled organisms (bacteria)—is essential to removing the dissolved BOD_5 from the solution and the removal of biomass from the system, as waste activated sludge, is essential to removing the BOD_5 from the system. Without the wasting of sludge, the BOD_5 would never actually leave the system—it would continually recycle internally.

3.5.1 FIXED FILM SYSTEMS

Figure 3.3 shows how slime builds up on a fixed film media—with an anaerobic zone at the interface, covered by an aerobic zone that the wastewater flows across. As the layers get thicker, eventually they become so thick that the nitrogen necessary to support the anaerobes cannot get through fast enough, the anaerobes die, and a large patch of growth breaks away—sloughs off—and goes to the secondary clarifier. From the clarifier it is removed from the system to complete the conversion cycle from dissolved BOD_5 to waste sludge.

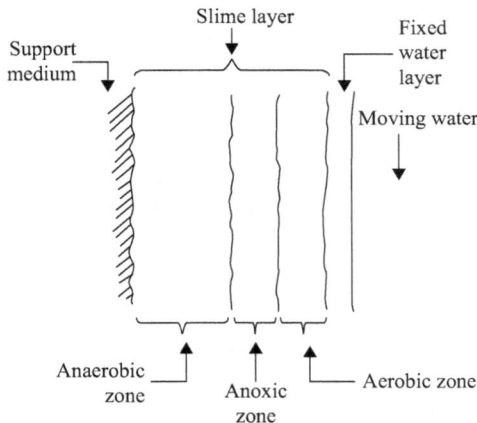

Figure 3.3. Slime build-up on a fixed film media.

3.5.1.1 Trickling Filters

The classic fixed film reactor is a trickling filter. In concept, the wastewater is trickled over an open bed of stone. The water runs through the interstitial spaces between the stones creating an opportunity for a biological slime to develop on the surfaces of the stones. As the water runs across the surface of the slime, the organisms in the slime absorb nutrients in the wastewater and use them for their own metabolic growth. See Figure 3.4 for a diagram of a trickling filter.

Functionally, the rotating arm depicted in Figure 3.4 distributes the wastewater across the surface of the media, the water runs through the stone where a slime layer builds up, the slime grabs the dissolved BOD_5 as the wastewater trickles over it, and a relatively clean effluent is discharged to the secondary clarifier. Trickling filters must be preceded by primary sedimentation to ovoid overloading the filter with larger solids.

In certain desert environments, trickling filters have been designed without walls, using only a wall of sand surrounding the bed of stone to hold everything together. In those cases, the water flows through the filter directly into the subsurface soils. Filters of this type need to be large enough to ensure total conversion of all of the organic material to carbon dioxide and water to avoid clogging of the pore spaces between the stones. Sometimes these systems will have underdrains through which the

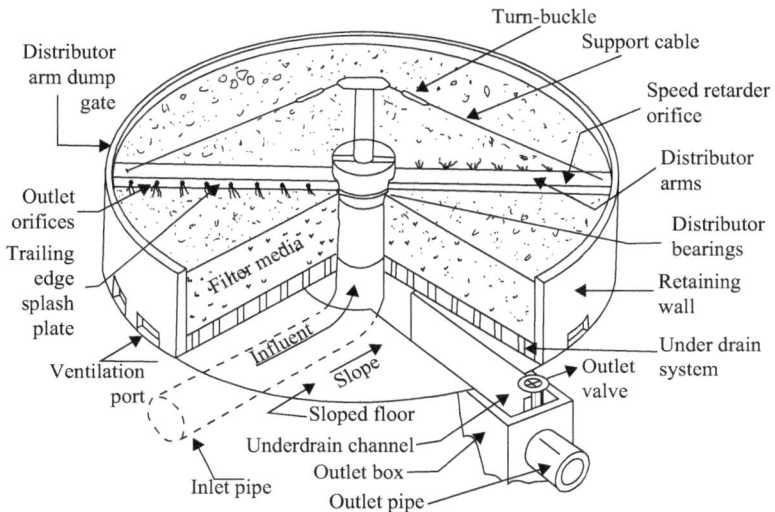

Figure 3.4. A classic trickling filter using a stone media bed (Adapted from: www.citywatertown.org/ and other places).

leachate runs directly into a seepage pit adjacent to the filter, rather than directly under the filter. Local conditions tend to dictate these differences.

Modern trickling filters are typically constructed using various plastic media, rather than a stone medium. The plastic media can take the form of large blocks of plastic sheets welded together and then cut to fit inside the round basin of the trickling filter. Other media include plastic balls or other shapes designed to provide huge surface areas for slime to grow on per cubic volume of media. Some plastic pellets have been designed for specialized applications that have as much as 11,000 sf of surface area per cubic foot of pellets (Bioquatic Supply Corporation). Most plastic media have surface areas of around 3,000 to 5,000 sf per cubic foot of volume. Units using media other than stone are generally referred to as "biological towers." They are further described in Section 3.6.1.2.

Table 3.8 shows the typical design parameters for a secondary treatment system, including those applicable to a trickling filter. Among those is the BOD_5 loading. This value relates the concentration of BOD_5 in the primary effluent to the surface area of the trickling filter media.

Equation 3.6 shows the calculation of BOD_5 loading on a trickling filter. That loading is based on the BOD_5 of the influent to the trickling filter from the primary sedimentation basin, without regard to the BOD_5 in any recycled wastewater. The hydraulic loading, on the other hand, does include both the influent to the filter and the recycle flow and it is calculated using Equation 3.7. The recycle ratio is calculated using Equation 3.8 and is equal to the recycled flow divided by the influent flow.

BOD_5 loading = (Primary effluent BOD_5)/(Volume of filter media) (3.6)

Where:

> BOD_5 loading = the pounds of BOD_5 applied to the filter per 1,000 cf of filter media per day (or in grams/m^3/d)
>
> Primary effluent BOD_5 = the pounds of BOD_5 applied per day, calculated from the concentration of BOD_5 in the primary effluent in mg/L times the average daily flow volume in mgd times 8.34 (or grams of BOD_5 applied per day)
>
> Volume of filter media = the volume of the stone, or other filter media, in thousands of cubic feet (or in cubic meters)

BOD_5 loading is also calculated in terms of pounds of BOD_5 per acre-foot of media per day. That equation is the following:

$$BOD_5 \text{ loading} = \text{(the pounds of } BOD_5 \text{ applied per day)/} \\ \text{(Volume of media in acre-ft)} \quad (3.7)$$

One acre-foot is equal to a surface area of one acre times a depth of one foot, or 43,560 cubic feet of media.

$$\text{Hydraulic loading} = (Q + Q_R)/A \qquad (3.8)$$

Where:

Hydraulic loading = the overflow rate on the filter surface in millions of gallons per acre of filter surface (or cubic meters per square meter per day)

Q = wastewater flow rate in millions of gallons per day (or in cubic meters per day)

Q_R = recirculation flow rate in millions of gallons per day (or in cubic meters per day)

A = the surface area of the filter in acres (or in square meters)

The recirculation flow rate is often specified in terms of a percentage of the incoming wastewater flow rate for convenience. That equation is the following.

$$R = Q_R/Q \qquad (3.9)$$

Where:

R = Recirculation ratio as a percentage, and the other terms, are as defined earlier

Table 3.8 shows typical design loadings for high rate and low rate trickling filters. Consistent with the data in that table, high rate filters require a much cleaner influent than normal primary sedimentation effluent.

A lot of equations have been developed over the years to calculate the efficiency of BOD_5 removal in trickling filters—most have, unfortunately, proven to not be very accurate. The National Research Council did some work in the 1940s, however, which turned out to be very good and is the basis for most of the current design standards for these types of plants. There are many variations of these equations in existence. Four of the sets of equations are shown in Table 3.9. Note that three of the equations were developed for metric units and use the BOD_5 loading in kg/m^3 of filter volume/day, while the first set of equations uses U.S. Standard units in lbs. of $BOD_5/1,000$ ft^3 of filter volume/day. Table 3.10 shows the calculation of efficiency based on the various equations. The first stage efficiencies are consistent between the U.S. Standard equations from Hammer and Hammer and the metric equations from Droste. The second stage efficiencies are drastically different depending on the equation used. It is noted that the McGhee equations and the Droste equations have good correlation in both the first stage and second stage efficiency values. The McGhee first stage values are slightly more conservative than either the Hammer and Hammer values or the Droste values.

Table 3.8. Typical secondary system design parameters

Process	Lbs. of BOD$_5$ loading per cubic foot per day (kg/m³/d)	Lbs. of BOD$_5$ loading per lb. MLVSS per day (kg/kg/d)	Mean cell residence time (sludge age), days	Recirculation ratio	Air supplied, in cf of air per lb. of BOD$_5$ (m³/kg)	Hydraulic loading rate (including recirculation flow) in ft³/ft² of top surface area/day (m³/ m²/d)	Organic loading rate in lbs. of BOD$_5$ per cubic foot of filter volume per day (kg/ m³/d)
Complete mix activated sludge			3–15	0.25–1.0			
Conventional activated sludge	0.56 (8.99)	0.2–0.5	3–15	0.25–0.70	45–90 (2.79–5.58)		
Extended aeration	0.32 (5.1)	0.05–0.20	20–40	0.5–1.5	90–125 (5.58–7.75)		
Tapered aeration	0.56 (8.99)	0.2–0.5	4–14	0.15–0.30	45–90 (2.79–5.58)		
Step feed	0.8 (12.8)	0.2–0.5	3–15	0.2–0.75	45–90 (2.79–5.58)		
Contact stabilization	1.12 (17.98)	0.2–0.5	5–10	0.5–1.5	45–90 (2.79–5.58)		
Oxidation ditch			15–30	0.75–1.5			

Sequencing batch reactors			10–30			
High rate activated sludge	1.6–6.4 (25.68–102.7)	0.5–3.5	0.8–4	25–45 (1.55–2.79)		
Conventional trickling filters				0.5–3.0	6–12 (1.8–3.7)	0.3–1.5 (4.8–24.1)
High rate trickling filters				1.0–5.0	30–90 (9.15–27.45)	1.5–18.7 (24.1–300)
Very high rate trickling filters				5.0–10.0	60–90 (18.3–27.45)	
Roughing filter				Generally not done	No limit—needs further treatment	No limit—needs further treatment
Rotating biological contactors (secondary treatment only)			Hydraulic retention time 0.7–1.5 hour		Based on exposed media surface area 0.27–0.54 (0.08–0.16)	Lbs. per 1,000 sf (kg per 100 m²) of exposed media surface area per day 2–3.5 (10.52–18.4)

(Continued)

Table 3.8. (*Continued*)

Process	Lbs. of BOD$_5$ loading per cubic foot per day (kg/m³/d)	Lbs. of BOD$_5$ loading per lb. MLVSS per day (kg/kg/d)	Mean cell residence time (sludge age), days	Recirculation ratio	Air supplied, in cf of air per lb. of BOD$_5$ (m³/kg)	Hydraulic loading rate (including recirculation flow) in ft³/ft² of top surface area/day (m³/m²/d)	Organic loading rate in lbs. of BOD$_5$ per cubic foot of filter volume per day (kg/m³/d)
Rotating biological contactors (combined with nitrification)			Hydraulic retention time 1.5–4 hours			Based on exposed media surface area 0.10–0.27 (0.03–0.08)	Lbs. per 1,000 sf (kg per 100 m²) of exposed media surface area per day 1.5–3 (7.9–15.8)
Rotating biological contactors (for nitrification, only)			Hydraulic retention time 1.2–2.9 hours			Based on exposed media surface area 0.13–0.34 (0.04–0.10)	Lbs. per 1,000 sf (kg per 100 m²) of exposed media surface area per day 0.2–0.6 (1.1–3.2)

Table 3.9. Comparison of NRC equation variations for trickling filter design

Definitions of terms	E_1 = BOD$_5$ removal efficiency in first stage filter, in %
	E_2 = BOD$_5$ removal efficiency in second stage filter, in %
	W = BOD$_5$ loading on first stage filter in lbs/1,000 ft³/day or kg/m³/day
	W' = BOD$_5$ loading on first stage filter in lbs/1,000 ft³/day or kg/m³/day
	V = Volume of the filter stage in 1,000 ft³ or m³
	F = Recirculation factor calculated per Equation 3.11

Hammer and Hammer (5th Edition)
W' = BOD$_5$ loading per 1,000 cf on the second stage unit

$$E_1 = 100/(1+(0.0561)\sqrt{(W/VF)}$$
$$E_2 = 100/(1+(0.0561)\sqrt{(W'/VF)}$$

McGhee (6th edition)
Note: Q = flow rate; C_i = influent BOD$_5$ concentration; C_e = the first stage effluent BOD$_5$ concentration; C_e' = second stage effluent BOD$_5$ concentration
Volumes and recirculation factors are for the specific reactor

$$(C_i - C_e)/C_i = 1/(1+(0.532))\,\sqrt{(QCi/VF)}$$
$$(C_e - C_e')/C_e = 1/(1+(\frac{0.532}{(1-E_1)}))\,\sqrt{(QCe/VF)}$$

Droste (1st Edition)

$$E_1 = 100/(1+ (0.443)\,\sqrt{(W/VF)}$$
$$E_2 = 100/(1+ (\frac{0.443}{(1-E_1)}))\,\sqrt{(W'/VF)}$$

Davis (1st Edition)
Note: Q = flow rate; C_i = influent BOD$_5$ concentration; C_e = the first stage effluent BOD$_5$ concentration
Volumes and recirculation factors are for the specific reactor

$$E_2 = 1/(1 + (4.12\,\sqrt{QC_i/VF})$$
$$E_2 = 1/(1 + (\frac{4.12}{(1-E_1)}\,\sqrt{QCe/VF})$$

Predictably, these equations give differing results for the efficiency of the two stages of a trickling filter system. The results, based on a filter volume (V) of 1,000 cubic feet or 1 cubic meter, a recirculation factor (F) as stated (calculated per Equation 3.11), and a BOD_5 loading in lbs/1,000 ft³ or kg/m³, as shown, are provided in Table 3.10.

It is noted that the various equations do provide reasonably consistent results, in most cases.

The NRC equation for filter efficiency for the first stage of a 2-stage filter system is the following:

$$E_1 = (100)/[1 + 0.056 \sqrt{(W/VF)}] \tag{3.10}$$

Where:

E_1 = First stage BOD_5 removal efficiency, in %
W = BOD_5 loading rate on first stage filter, in lbs/day
V = Volume of filter media, in 10^3 cf
F = Recirculation factor

The recirculation factor is calculated from:

$$F = (1 + R)/(1 + 0.1R)^2 \tag{3.11}$$

Where:

R = Recirculation ratio (Recirculated flow/Influent flow),
                    both in gals/day $\tag{3.12}$

The second stage efficiency is calculated using the NRC equations as follows:

$$E_2 = (100)/\{1 + [0.0561/(1 - E_1)] [\sqrt{(W/VF)}]\} \tag{3.13}$$

Where:

W' = BOD_5 loading to the second stage filter, in lbs/day

$$W' = (1 - E_1) W \tag{3.14}$$

The efficiency of both stages is affected by the temperature of the wastewater. The equations are based on water temperatures of 20°C. For temperatures other than 20°C, a correction factor is applied, as follows:

$$E_t = E_{20} \Theta^{(t-20)} \tag{3.15}$$

Table 3.10. First and second stage trickling filter efficiencies for BOD$_5$ loadings in lbs/1,000 cf/day or kg/m³/day for reactors of 1,000 cf or 1 m³ and the recirculation factors shown

W	V	R	F	W'	E$_1$ Hammer and Hammer	E$_2$ Hammer and Hammer	E$_1$ McGhee[1]	E$_2$ McGhee[1]	E$_1$ Droste	E$_2$ Droste	E$_1$ Davis[1,2]	E$_2$ Davis[1,2]
10	1	0	1.00	50	84.9	71.6	82.4	45.2	84.9	45.9	37.7	84.9
10	1	1	1.65	50	87.9	76.4	85.8	46.2	87.9	46.8	43.7	87.8
10	1	2	2.08	50	89.1	78.4	87.1	46.6	89.0	47.1	46.6	89.0
10	1	3	2.37	50	89.7	79.5	87.8	46.8	89.6	47.3	48.2	89.6
20	1	0	1.00	100	79.9	64.1	76.8	43.4	79.9	44.4	30.0	79.9
20	1	1	1.65	100	83.7	69.6	81.0	44.7	83.6	45.5	35.5	83.6
20	1	2	2.08	100	85.2	72.0	82.7	45.3	85.2	46.0	38.2	85.1
20	1	3	2.37	100	86.0	73.3	83.6	45.5	86.0	46.2	39.7	85.9
40	1	0	1.00	200	73.8	55.8	70.1	41.2	73.8	42.5	23.2	73.7
40	1	1	1.65	200	78.4	61.8	75.1	42.9	78.3	43.9	28.0	78.3
40	1	2	2.08	200	80.3	64.5	77.2	43.6	80.2	44.5	30.4	80.2
40	1	3	2.37	200	81.3	66.0	78.3	43.9	81.2	44.8	31.8	81.2

(*Continued*)

Table 3.10. (*Continued*)

W	V	R	F	W'	E₁ Hammer and Hammer	E₂ Hammer and Hammer	E₁ McGhee[1]	E₂ McGhee[1]	E₁ Droste	E₂ Droste	E₁ Davis[1,2]	E₂ Davis[1,2]
60	1	0	1.00	300	69.7	50.7	65.7	39.6	69.7	41.1	19.8	69.6
60	1	1	1.65	300	74.7	57.0	71.1	41.6	74.7	42.8	24.1	74.7
60	1	2	2.08	300	76.9	59.8	73.4	42.3	76.8	43.4	26.3	76.8
60	1	3	2.37	300	78.0	61.3	74.6	42.7	77.9	43.8	27.5	77.9
80	1	0	1.00	400	66.6	47.1	62.4	38.4	66.5	40.0	17.6	66.5
80	1	1	1.65	400	71.9	53.4	68.0	40.5	71.9	41.8	21.6	71.8
80	1	2	2.08	400	74.2	56.3	70.5	41.4	74.2	42.6	23.6	74.1
80	1	3	2.37	400	75.4	57.8	71.8	41.8	75.4	43.0	24.8	75.3
100	1	0	1.00	500	64.1	44.4	59.7	37.4	64.0	39.0	16.1	64.0
100	1	1	1.65	500	69.6	50.6	65.6	39.6	69.6	41.0	19.7	69.5
100	1	2	2.08	500	72.0	53.5	68.1	40.5	72.0	41.9	21.6	71.9
100	1	3	2.37	500	73.3	55.1	69.5	41.0	73.2	42.3	22.7	73.2
120	1	0	1.00	600	61.9	42.1	57.5	36.5	61.9	38.2	14.9	61.9
120	1	1	1.65	600	67.7	48.3	63.5	38.8	67.6	40.3	18.3	67.6
120	1	2	2.08	600	70.1	51.2	66.1	39.8	70.1	41.2	20.1	70.1

120	1	3	2.37	600	71.5	52.8	67.5	40.3	71.4	41.7	21.2	71.4
150	1	0	1.00	700	59.3	40.3	54.7	35.4	59.2	37.2	13.5	59.2
150	1	1	1.65	700	65.2	46.4	60.9	37.8	65.1	39.4	16.7	65.1
150	1	2	2.08	700	67.7	49.3	63.6	38.9	67.7	40.4	18.4	67.7
150	1	3	2.37	700	69.1	50.9	65.0	39.4	69.1	40.9	19.4	69.0
200	1	0	1.00	800	55.8	38.7	51.2	33.8	55.7	35.8	11.9	55.7
200	1	1	1.65	800	61.8	44.8	57.4	36.5	61.8	38.2	14.8	61.8
200	1	2	2.08	800	64.5	47.6	60.2	37.6	64.5	39.2	16.3	64.4
200	1	3	2.37	800	66.0	49.2	61.7	38.2	65.9	39.7	17.2	65.9

[1] Here W and W' are substituted for QC_i and QC_e and 100 is used for the numerator to convert to efficiency.

[2] The Davis equations seem to reverse the efficiencies of the first and second stages. That may be due to the use of the second stage effluent BOD_5 concentration in the Davis second stage equation. That complicates the conversions done here and these calculations may reflect inaccurate conversions from the Davis equations to comparable values from the other sources. Caution should be used when using the values in this table for the Davis equations, although direct use is expected to provide favorable outputs.

Where:

Θ = 1.035 for most practical applications
E_t = Efficiency at desired temperature
E_{20} = Efficiency calculated for 20°C

Figures 3.5 and 3.6 provide a graphic representation of the data developed from the Droste equations, as calculated in the table earlier. By entering Figure 3.5 with a BOD_5 loading per 1,000 cf of filter volume and a desired removal efficiency in the first stage or in a single stage reactor, the required recycle rate can be closely estimated. The higher the recirculation ratio, the larger the resulting filter has to be because of the limitations on the surface overflow rates that have been empirically determined by the NRC to be most effective. Figure 3.6 will provide similar results for the second stage filter. The input data to the second stage are assumed to have been adjusted for the reduced influent BOD_5 loading on the secondary filter based on Equation 3.10.

The efficiency necessary to directly achieve secondary discharge standards is not realistically feasible with a single stage system unless the primary sedimentation basin is extremely efficient or the influent BOD_5 to the treatment plant is extremely light. Therefore, an intermediate clarifier and a second stage filter are routinely required with this type of treatment

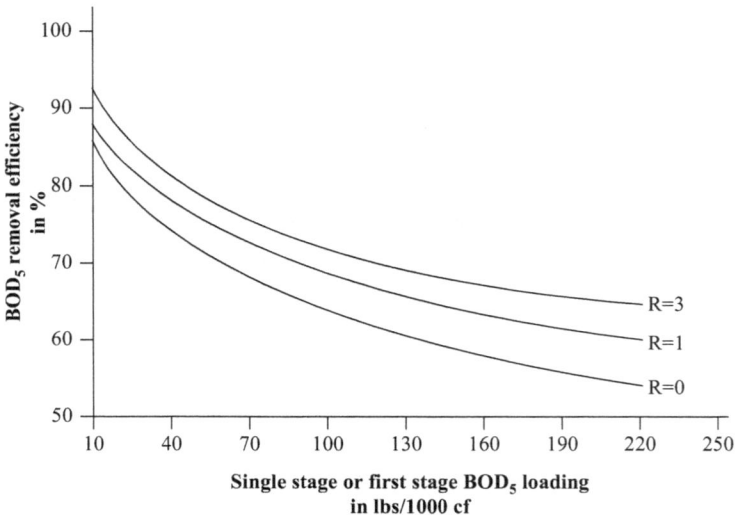

Figure 3.5. First stage or single stage BOD_5 loadings plotted against removal efficiency for various recirculation factors. Based on Droste equations.

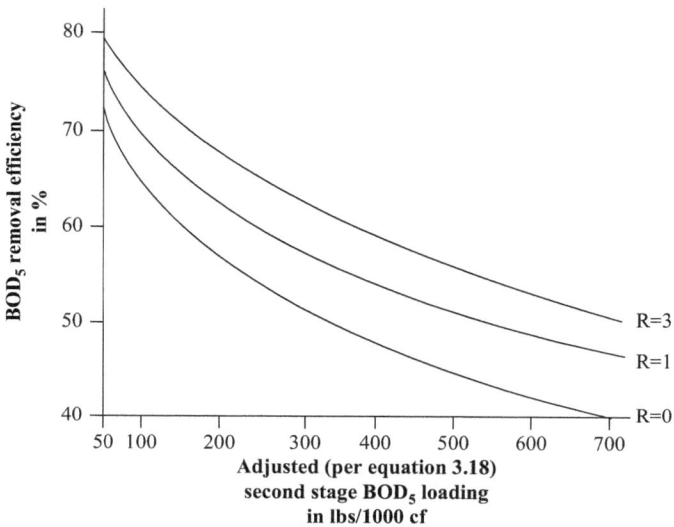

Figure 3.6. Second stage BOD_5 loadings plotted against removal efficiencies for various recirculation factors. Based on Droste equations and assumes that BOD_5 loadings have been corrected using Equation 3.18.

scheme. A single stage system is often used as a pretreatment for very strong wastewater prior to discharge to a conventional activated sludge secondary system. The intermediary clarifier is not always used when the filter is used in this fashion since the sludge will be incorporated into the secondary sludge mix later in any case. Some recirculation of filter effluent to the head of the filter is generally used in any case to achieve the desired removal efficiencies in the trickling filter.

If used, an intermediate clarifier is designed the same way a primary clarifier is designed, but the BOD_5 loading is reduced by the efficiency of the primary clarifier and the first stage trickling filter.

Figure 3.6 is based on a restatement of the BOD_5 strength of the first stage effluent to account for the fact that the most easily degraded materials are destroyed in the first stage and only the harder-to-deal-with materials are left for the second stage. The correction for that restatement is calculated from Equation 3.18.

An alternative approach to calculating or predicting the overall efficiency of a trickling filter system is the following.

(Second stage BOD_5 corrected for treatability) =
First stage effluent BOD_5 load/$(1 - \text{First stage efficiency})^2$ (3.16)

Where:

the BOD$_5$ load for both stages is expressed in pounds of BOD$_5$ applied per day, and the first stage efficiency is expressed as a decimal percentage.

The overall efficiency of the two-stage process is determined from Equation 3.17 in the same way the first stage efficiency was determined using the BOD$_5$ loading from the intermediate clarifier, adjusted for treatability using Equation 3.16.

$$E_T = 1 - [(1 - P_E)(1 - E_1)(1 - E_2)] \qquad (3.17)$$

Where:

E_T = Total plant removal efficiency, as a decimal percentage
P_E = Primary sedimentation basin removal efficiency, as a decimal percentage
E_1 = Removal efficiency of the first stage filter, as a decimal percentage
E_2 = Removal efficiency of the second stage filter, as a decimal percentage

The required efficiency of the second stage filter can be calculated from the following equation:

$$E_2 = 1 - [E_T/(1 - P_E)(1 - E_1)] \qquad (3.18)$$

Where:

E_2 = Required efficiency of the second stage filter, decimal percentage
E_T = Total desired efficiency of the treatment system, decimal percentage
P_E = Efficiency of the primary sedimentation basin, decimal percentage
E_1 = Efficiency of the first stage filter, decimal percentage

The efficiency of a trickling filter is strongly influenced by temperature, as shown by Equation 3.19, which is identical to Equation 3.15 presented earlier.

$$E_t = (E_{20})(1.035)^{(T-20)} \qquad (3.19)$$

Where:

E_t = Removal efficiency at the actual reactor operating temperature, °C
E_{20} = Total removal efficiency calculated from Equation 3.11
T = Actual reactor operating temperature, degrees C

Example Problems 3.4 and 3.5 demonstrate the use of the NRC equations.

Example Problem 3.4

Given a municipal wastewater treatment facility with an influent BOD_5 of 300 mg/L and an effluent BOD_5 requirement of 25 mg/L, find the diameter of each stage of a two-stage trickling filter system if the flow is 3.0 mgd through the plant, the desired filter depths are 10 feet, and the desired recirculation ratio is 2:1. Assume that $E_1 = E_2$ and the average wastewater temperature is 20°C.

Solution

First, calculate the required value for E_1 and E_2.

The overall efficiency of the facility must be
$E_t = (300 \text{ mg/L} - 25 \text{ mg/L})/350 \text{ mg/L}$
$\quad = 92\%$
$E_1 + E_2 (1 - E_1) = 0.92$
$E_2 = E_1$

Therefore:

$2 E_1 - E_1{}^2 = 0.92$
$E_1{}^2 - 2 E_1 + 0.92 = 0$

And

$E_1 = E_2 = 71\%$

Then compute the recirculation factor using Equation 3.11:

$F = (1 + R)/(1 + 0.1R)^2$
$F = 3/(1.2)^2$
$F = 2.08$

Compute the BOD_5 loading for the first filter. Note that the conversion of mg/L to lbs/day is done empirically with the following equation:

$$\text{(mg/L) (flow in mgd) (8.34)} = \text{lbs/day} \qquad (3.20)$$

Therefore,

$$W = (300)\ (3)\ (8.34) = 7{,}506 \text{ lbs/day}$$

The volume for the first stage filter is then calculated using Equation 3.10 as:

$E_1 = (100)/[1 + 0.056\ \sqrt{(W/VF)}\ ]$
$71 = (100)/[1 + 0.056\ \sqrt{(7{,}506/(2.08\ V)}]$
$71 + 3.976\ \sqrt{(7{,}506/(2.08\ V)} = 100$
$\sqrt{(7{,}506/(2.08\ V)} = 7.29$
$7506 = 110.65\ V$
$V = 67.8 \times 10^3\ \text{ft}^3$
$67.8 \times 10^3\ \text{ft}^3 = (10)\ (\pi)\ (d)^2/4$
$(d)^2 = 8{,}632.6\ \text{ft}^2$
$d = 92.9\ \text{ft}$

Calculate the BOD_5 loading on the second stage using Equation 3.14 as follows:

$W' = (1 - 0.71)\ W$
$W' = (0.29)\ (7{,}506) = 2{,}177$

Calculate the required volume of the second filter from Equation 3.13 as follows:

$E_2 = (100)/\{1 + [0.0561/(1 - E_1)]\ [\sqrt{(W'/VF)}]\}$
$71 = (100)/\{1 + [0.0561/(1 - 0.71)]\ [\sqrt{(2{,}177/2.08\ V)}]\}$
$71 = (100)/[1 + 0.193\ \sqrt{(2{,}177/(2.08\ V)}]$
$71 + 13.71\ \sqrt{(2{,}177/(2.08\ V)} = 100$
$\sqrt{(2{,}177/\ (2.08\ V)} = 2.12$
$2{,}177 = 9.31\ V$
$V = 233.8 \times 10^3\ \text{ft}^3$
$233.8 \times 10^3\ \text{ft}^3 = (10)\ (\pi)\ (d)^2/4$
$(d)^2 = 29{,}768\ \text{ft}^2$
$d = 172.5\ \text{ft}$

That is very large for a filter, so two smaller filters of 122 ft diameter may be more practical. Note that diameters are usually rounded to the nearest five feet to accommodate standard rotary distributor systems. Here, that would mean a first stage diameter of 95 ft and a second stage diameter of 175 feet, or two at 125 ft.

Example Problem 3.5

Using the NRC equations for designing a two-stage trickling filter system, determine the BOD_5 loading (in lbs/day), the hydraulic loading on the filter (in ft/day), the expected BOD_5 removal efficiency for each stage, and the expected effluent BOD_5 concentration, given the following system characteristics:

Influent BOD_5 concentration = 150 mg/L
Recirculation rate = 100% in the first stage and 75% in the second stage
Filter diameter = 25 m
Average daily flow rate = 3.0 mgd
Depth of filter media = 2.0 m
Average wastewater temperature = 18°C

Solution

Calculate the BOD_5 loading on the first stage as follows:

$$(150 \text{ mg/L}) (3.0 \text{ mgd}) (8.34) = \underline{3,753 \text{ lbs of BOD5 per day}}$$

Calculate the hydraulic loading rate as follows:

$A = \pi (d)^2/4$
$A = \pi (25 \text{ m})^2/4$
$A = 490.9 \text{ m}^2$
$490.9 \text{ m}^2 \times (3.281)^2 \text{ ft}^2/\text{m}^2 = 5,284.2 \text{ ft}^2$
$Q = (3 \times 10^6 \text{ gal/day}) \times (1 \text{ ft}^3/7.48 \text{ gal}) = 401,070 \text{ ft}^3/\text{day}$
Hydraulic loading rate = $Q/A = (401,070 \text{ ft}^3/\text{day})/(5,284.2 \text{ ft}^2)$
   = 75.9 ft/day

Calculate the first stage removal efficiency using Equation 3.10 as follows:

$F = (1 + R)/(1 + 0.1R)^2$
$F = (2)/(1.1)^2 = 1.65$
$E_1 = (100)/[1 + 0.0561 \sqrt{(W/VF)}]$
$W = 3,753$
$V = (490.9 \text{ m}^2) (2.0 \text{ m}) (3.281 \text{ ft/m})^3 = 34,667 \text{ ft}^3$
$F = 1.65$
$E_1 = (100)/[1 + 0.0561 \sqrt{(3,753/1.65 \times 34.7)}]$

$E_1 = 68.8\%$
Correct to 18°C as follows.
$E_{18} = (68.8)(1.035)^{-2} = \underline{64.2\%}$

Calculate the BOD_5 loading on the second stage as follows:

$$(150 \text{ mg/L})(1-0.642)(3.0 \text{ mgd})(8.34) = 1{,}343.6 \text{ lbs/day}$$

Calculate the second stage removal efficiency as follows:

$$E_2 = (100)/\{1 + [0.0561/(1 - E_1)][\sqrt{(W'/VF)}]\}$$

Where:

$E_1 = 83.7$
$W' = 1{,}343.6 \text{ lbs/day}$
$F = (1.75)/(1.075)^2 = 1.514$
$V_2 = V_1 = 34{,}667 \text{ ft}^3$

Then:

$$E_2 = (100)/\{1 + [0.0561/(.358)][\sqrt{(1{,}343.6/34.7 \times 1.514)}]\}$$
$$E_2 = \underline{55.8\%}$$

Correcting to 18°C:

$E_{18} = (55.8)(1.035)^{-2}$
$E_{18} = 52.1\%$

Calculate the expected effluent BOD_5 as:

$$(150 \text{ mg/L})(1-0.642)(1-0.521) = \underline{25.7 \text{ mg/L}}$$

With trickling filters, the amount of sludge generated for direct disposal is generally very small. That is due to the recycling of the filter effluent back to the head of the filter and the recycling of settled sludge in the intermediate and final clarifiers to the head of the treatment plant. Most of the organic fractions of the sludge from the secondary portion of these types of treatment systems are converted to CO_2 and water, with a small fraction of inorganic constituents remaining for removal through the primary sedimentation basin sludge removal system.

Table 3.11. Trickling filter design and performance parameters

	Hydraulic loading, in ft/day (m/day)	Organic loading, in lbs. of BOD_5/d per 1,000 ft³ of media (g/m³ per d)	Organic loading for secondary treatment, in lbs. of BOD_5/d per ft³ of media (kg/d per m³)	Organic loading for BOD_5 removal plus nitrification, in lbs. of BOD_5/d per ft³ of media (g/d per m³)	Nitrogen loading for tertiary nitrification, in lbs. of NH_4-N/d per ft² of media (g/d per m²)	Recirculation ratio	Depth, in feet (m)	BOD_5 removal efficiency, in %
Roughing filter prior to secondary	200–600 (60–180) plus recirculation	>60 (>1,000)	0.1–0.25 (1.6–4)			0–2	3–20 (1–6)	40–85
Standard (low) rate filter	3–12 (1–4)	5–15 (80–240)				0	5–10 (1.5–3)	80–85
Intermediate rate filter	12–35 (4–10)	15–30 (250–500)				0–1	5–8 (1.5–2.5)	50–70
High rate filter (stone media)	35–130 (10–40)	20–110 (320–1,800)				0.5–3	3–7 (1–2)	40–80

(Continued)

Table 3.11. (*Continued*)

	Hydraulic loading, in ft/day (m/day)	Organic loading, in lbs. of BOD₅/d per 1,000 ft³ of media (g/m³ per d)	Organic loading for secondary treatment, in lbs. of BOD₅/d per ft³ of media (kg/d per m³)	Organic loading for BOD₅ removal plus nitrification, in lbs. of BOD₅/d per ft³ of media (g/d per m³)	Nitrogen loading for tertiary nitrification, in lbs. of NH₄-N/d per ft² of media (g/d per m²)	Recirculation ratio	Depth, in feet (m)	BOD₅ removal efficiency, in %
High rate filter (stone media) two-stage		45–70 (725–1,120)				0.5–4	5–7 (1.5–2)	75–85
Super high rate (plastic media)	50–300 (15–90) plus recirculation	20–60 (320–960)	0.02–0.06 (0.3–1)	0.006–0.019 (0.1–0.3 kg/d per m³) [0.000012–0.00006 (0.2–1 g TKN/ m²-d)]	0.00003–0.00016 (0.5–2.5 g NH₄-N/m²-d	1–2	Up to 40 (Up to 12)	65–85
Super high rate (plastic media) effluent quality			15–30 mg/L BOD₅ 15–30 mg/L TSS	<10 mg/L BOD₅ <3 mg/L NH₄-N	0.5–3 mg/L NH₄-N			

Source: Davis (2011); Metcalf & Eddy (2014)

3.5.1.2 Biological Towers

Various types of media are used in trickling filters to achieve BOD_5 removals. When media other than river run stone is used, the resulting reactor is generally referred to as a biological tower. The materials used are often a "random packing," a loose accumulation of variously shaped plastic beads, balls, and amorphous beads. The plastic shapes generally provide about 30 to 40 square feet of surface area for biological growth per cubic foot of material (100 to 130 m^2/m^3). The void ratios run in the order of 90 to 95 percent. Beads tend to offer much higher surface area per cubic foot of volume with values running in the 3,000 to 4,000 square feet per cubic foot with some specialty beads running as high as 11,000 square feet per cubic foot. Beads tend to require a much higher volume to fill a reactor and they tend not to drain as well since the void ratio is also much less than that of plastic shapes. Consequently, hydraulic loading rates tend to be proportional to the void ratio and are smaller for the beads.

Other common materials for towers are blocks of corrugated sheets welded to intermediary flat sheets. Some of these provide vertical channels for water flow and for slime growth. Others provide cross-directional channels for the same purpose. These blocks tend to offer surface area per unit volume and void ratio numbers consistent with the random shapes. Manufacturers should be contacted for specific data at the time of use due to the constant redevelopment of the shapes and changing values for those parameters.

The equations for these kinds of media are based on first order reaction kinetics. They are as follows. These equations are based on the removal of soluble BOD_5, since the residence time inside a tower is insufficient for significant solubilization to occur, and if the towers are dosed with a high concentration of nonsolubilized BOD_5 they are likely to clog. To remove as much of the suspended solids as reasonably possible before discharge to a packed tower, a sedimentation process or combination of comminutor and fine screening is required.

Removal efficiency of towers is typically calculated from the following equation, adapted from Hammer and Hammer (2008):

$$\text{Effluent } BOD_5/\text{Influent } BOD_5 = e^{(-k_{20}A_sD/Q^n)} \qquad (3.21)$$

Where:

k_{20} = Reaction rate coefficient at 20°C, in $(gal/min*ft^2)^{0.5}$ or $(L/m^2 * sec)^{0.5}$
A_s = Specific surface area of the filter media, in (ft^2/ft^3) or (m^2/m^3)
D = Depth of filter media, in ft or m

Q = Surface overflow rate, in (gal/min*ft²) or (L/m²*sec)
n = Flow constant for specific filter media, usually assumed as 0.5 for vertical flow and cross flow filter media

It is noted that the influent BOD_5 is often diluted with clear recirculation flow to reduce the organic loading on the filters. The equations for calculating the diluted influent BOD_5 value for use in Equation 3.22 are the following. It is noted that a "clear" recirculation flow does not imply a flow free of solubilized BOD_5. This term implies only a flow with very low suspended solids concentration. It may still contain to a significant solubilized BOD_5 concentration:

$$R = Q_r/Q_p \qquad (3.22)$$

Where:

Q_r = Surface overflow rate of recycled flow, in gal/min*ft² or L/min*m²
Q_p = Surface overflow rate of primary effluent without recirculation flow, in gal/min*ft² or L/min*m²

Then:

$$S_i = (S_p + RS_e)/(1 + R) \qquad (3.23)$$

Where:

S_i = Diluted influent BOD_5, in mg/L
S_p = Undiluted influent BOD_5, in mg/L
S_e = BOD_5 in filter effluent, in mg/L

Example Problem 3.6 illustrates the use of these equations.

Example Problem 3.6

A biological tower has the following characteristics:

Diameter = 30 ft
Depth of media = 30 ft
k_{20} = 0.0015 gal/min*ft²
Specific surface area = 40 ft²/ft³
n = 0.50

Influent wastewater is primary effluent

$Q_p = 0.65$ mgd
$S_p = 172$ mg/L
Recycled wastewater flow $BOD_5 = 85$ mg/L
$R = 1.0$

Calculate the effluent BOD_5 (S_e) in mg/L, assuming no recycle, and calculate the overall BOD_5 removal efficiency of the unit.

Solution

From Equation 3.23:

Effluent $BOD_5 =$ Influent BOD_5 (e $^{(-k_{20}A_s D/Q^n)}$)
$\qquad = (172$ mg/L) (e $^{(-0.0015\,(40)\,(30)/(0.65)\,^{0.5})}$)
$\qquad = (172$ mg/L) (e $^{(-1.8/0.806)}$)
$\qquad = (172$ mg/L) (0.107)
$\qquad = 18.4$ mg/L

Efficiency $= 1 - (18.4$ mg/L $/172$ mg/L)
$\qquad = 1 - 0.11$
$\qquad = 0.89$ or 89% BOD_5 removal efficiency

Example Problem 3.7

Assume the same parameters as for Example Problem 3.6, but recalculate the influent BOD_5 based on dilution from the recycled flows, then recalculate the effluent BOD_5 and the overall BOD_5 removal efficiency.

Solution

From Equation 3.25

$\qquad S_i = (S_p + RS_e)/(1 + R)$
S_i mg/L $= [(172$ mg/L $+ (1)\,(18.4$ mg/L)]/(1 + 1)
$\qquad = (190.4$ mg/L/2)
$\qquad = 95.2$ mg/L

Then:

$$\text{Effluent BOD}_5 = \text{Influent BOD}_5 \, (e^{(-k_{20}A_s D/Q^n)})$$
$$= (95.2 \text{ mg/L}) \, (e^{(-0.0015\,(40)\,(30)/(0.65)\,^{\wedge}0.5)})$$
$$= (95.2 \text{ mg/L}) \, (e^{(-1.8/0.806)})$$
$$= (95.2 \text{ mg/L}) \, (0.107)$$
$$= 10.2 \text{ mg/L}$$

$$\text{Efficiency} = 1 - (10.2 \text{ mg/L} \, / 172 \text{ mg/L})$$
$$= 1 - 0.06$$
$$= 0.94 \text{ or } 94\% \text{ BOD}_5 \text{ removal efficiency}$$

Notice that the value of k_{20} in these equations is just as dependent upon temperature as for any other calculation using that term. The correction factor for k_{20} to any other temperature is the following:

$$k_T = (k_{20}) \, (\theta^{T-20}) \qquad\qquad (3.24)$$

Where:

k_T = Reaction rate at desired temperature
θ = Temperature correction coefficient, normally assumed to be 1.035
k_{20} = Reaction rate at 20°C
T = Temperature at which the value of k is desired, in °C

3.5.1.3 Rotating Biological Contactors

A previously common fixed film reactor that seems to be losing favor of late is the rotating biological contactor, or rotating biological contactors (RBCs) unit. These units are constructed of bundles of plastic sheets attached to a shaft. The shaft rotates over a rounded bottom tank with a portion of the plastic submerged at all times. The concept behind a RBC unit is to provide a constantly wet surface for a biological slime to grow on while minimizing the amount of air that needs to be supplied to the wastewater to maintain aerobic conditions. By rotating large bundles of plastic sheets into and out of the wastewater on a regular, set time basis, wastewater slime can accumulate on the surface of the plastic disks. When a portion of the disk is submerged, the slime absorbs nutrients from the wastewater for use in biological growth. When that area of the disk is out of the water, the slime absorbs oxygen from the surrounding atmosphere and assimilates the nutrients into new plant growth that is ready to receive new nutrients when the disk re-enters the wastewater.

RBCs typically have about 40 percent of the depth immersed in the wastewater at any time, but can be greater than 50 percent submerged if good aeration of the wastewater can be assured. The addition of the needed aeration capacity to maintain submergence greater than about 40 percent is typically so high for operating costs that the lower submergence and greater number of disks is more cost-effective over time. If the dissolved oxygen in the wastewater gets too low and a portion of the disk is always submerged, the biological growth on that portion of the disk can become anaerobic, causing odors and potential operational problems with sludge management. Therefore, if submergence is deep, aeration is added to enhance the aeration of the biological film and to assist with rotation of the bundles. Air cups are provided on the outside perimeter of the bundles by some manufacturers to assist with the bundle rotation, thereby saving energy costs associated with turning the massive bundles.

3.5.2 SUSPENDED GROWTH BIOLOGICAL TREATMENT SYSTEMS

With suspended growth systems, the biological activity occurs in a continuously stirred reactor rather than attached to a fixed film. There are several types or variations of suspended growth reactors including several variations of conventional activated sludge systems, plus lagoons, ponds, carousels, and other esoteric systems.

3.5.2.1 Conventional Activated Sludge Systems

A conventional activated sludge process is defined by several parameters. Among those parameters are the aeration period, the BOD_5 and total suspended solids (TSS) loadings per unit volume of the reactors, the food to microorganism (F/M) ratio, and the sludge age. Aeration period is calculated the same as detention time, using Equation 3.3. In essence, so long as the bacteria are in the aeration basin they are being aerated and when they leave they stop being aerated—thus the aeration period and detention time are identical.

BOD_5 and TSS loadings are usually expressed in terms of pounds of BOD_5 or TSS applied per thousand cubic feet of liquid volume in the reactor, using Equation 3.6 provided in the discussion of trickling filters. In this case, V is the liquid volume in the reactor, rather than the volume of stone in the trickling filter. Both parameters are calculated using the same equation since both are measured in terms of mg/L of concentration.

3.5.2.1.1 F/M Ratio

The F/M ratio is the ratio of the pounds of BOD_5 applied to the reactor per day per pound of suspended solids (the microbial mass) in the aeration basin liquid, known as the "mixed liquor."

This then is the pounds of BOD_5 applied to the reactor per day per pound of mixed liquor suspended solids (MLSS). This ratio is also sometimes calculated using the same symbols, but using the mixed liquor volatile suspended solids (MLVSS). This calculation is used because there is a significant school of thought that points out that only the organic solids are subject to biological degradation and that the volatile suspended solids represent the organic fraction. Therefore, using the MLVSS for this calculation, rather than the MLSS, provides a more realistic value of the actual F/M ratio. In those cases where the fixed suspended solids, or the inorganic fraction of the suspended solids, are abnormally high, this may be true. For most applications, however, the high variability of the influent wastewater characteristics is sufficient to render this refinement of limited value, particularly since most treatment plants maintain a consistent MLVSS:MLSS ratio. The F/M ratio is the key control parameter for operation of the activated sludge system, however, and must be calculated carefully. Equation 3.25 shows the calculation of the F/M ratio.

$$F/M = [(Q)\ (BOD_5)]/[(V)\ (MLSS)] \qquad (3.25)$$

Where:

F/M = Food to microorganism ratio in pounds of BOD_5 applied per day per pound of MLSS (kilograms of BOD_5 applied per day per kilogram of MLSS)

Q = Wastewater flow rate in millions of gallons per day (cubic meters per day)

BOD_5 = Concentration of BOD_5 in the influent wastewater in mg/L (grams/m³)

V = Liquid volume in the aeration basin in millions of gallons (cubic meters)

MLSS = Mixed liquor suspended solids concentration in mg/L (grams/m³)

There is a direct, inverse, relationship between the F/M ratio and the parameters of BOD_5 loading per unit volume per day and the aeration period. If the aeration period is changed, the BOD_5 loading is also changed in inverse proportion. If the aeration period is cut in half, for example, the BOD_5 loading per unit volume per day is doubled for the same BOD_5

concentration in the wastewater. On the other hand, the F/M ratio expresses the BOD_5 concentration as a function of the growth rate and concentration of the microbes in the reactor rather than as a function of the reactor volume. That means that the F/M ratio is not a function of either the basin volume or the aeration period. Thus, two different reactors, treating a wastewater with the same influent BOD_5 concentration but using very different aeration periods and reactor volumes, can have identical F/M ratios. A shorter aeration period with a higher MLSS concentration will yield the same F/M ratio as a longer aeration period with a lower MLSS concentration in a conventional activated sludge treatment plant.

3.5.2.1.2   Sludge Age

Sludge age, or mean cell residence time, is related to the F/M ratio. The concept here is that the wastewater goes through the aeration process only once, but the sludge (solids) go through several times due to the sludge recycle process. Aeration time in a conventional activated sludge system is in the order of 3 to 30 hours; the sludge age is typically in the order of several days. Equation 3.26 shows the calculation of sludge age in an activated sludge system.

$$\text{Sludge Age} = [(V)\,(MLSS)]/[(TSS_e)\,(Q_e) + (TSS_w)\,(Q_w)] \qquad (3.26)$$

Where:

> Sludge age = Mean cell residence time in days
> V = Liquid volume in the aeration basin in millions of gallons (cubic meters)
> MLSS = Mixed liquor suspended solids concentration in mg/L (grams/m³)
> TSS_e = Suspended solids concentration in the effluent of the reactor in mg/L (grams/m³)
> Q_e = Effluent flow rate in millions of gallons per day (cubic meters per day)
> TSS_w = Suspended solids concentration in the waste activated sludge (the sludge sent to the digesters) in mg/L (grams/m³)
> Q_w = Waste activated sludge flow rate in millions of gallons per day (cubic meters per day)

Sludge age can also be expressed in terms of volatile suspended solids concentrations, rather than total suspended solids. If the F/M ratio is being calculated using MLVSS then it is useful to maintain consistency and to use MLVSS for the sludge age calculations, as well.

Table 3.12. Typical loading and operational parameters for secondary treatment processes

	Detention time, in hours or days, as noted	BOD$_5$ loading, in lbs./day per 1,000 cf	F/M ratio, in lbs. of BOD$_5$/day per lb. of MLSS	MLSS, in mg/L	Sludge age, in days (sludge retention time)	Typical aeration period, in hours	Sludge recycle rate, without nutrient removal, in %	Typical BOD$_5$ removal efficiency, in %
Conventional activated sludge (complete mix)	3–15 hours	20–40	0.2–0.5 0.2–0.6 (MLVSS base)	1,000–4,000	3–15	4–7.5	20–100	80–90
Tapered aeration activated sludge	4–8 hours	40–60	0.2–0.5 0.2–0.4 (MLVSS base)	1,000–3,500	3–15	4–7	25–75	80–90
Step feed activated sludge	3–5 hours		0.2–0.5 0.2–0.4 (MLVSS base)	1,500–4,000	3–15		25–75	
Extended aeration activated sludge	20–30 hours	10–20	0.05–0.02 0.04–0.1 (MLVSS base)	2,000–8,000	20–40	20–30	50–100	85–95

High/pure oxygen activated sludge	1–3 hours	>120 (80–200, typical)	0.6–1.5 0.5–1 (MLVSS base)	2,000–8,000	1–10	1–3	30–50	80–90
High rate aeration	1–2 hours	75–150	1.5–2 (MLVSS base)	500–1,500	0.5–2	1–2		
Contact stabilization	0.5–1 hours in contact basin, 2–4 hours in stabilization basin	60–75	0.2–0.6 (MLVSS base)	1,000–3,000 in contact basin, 6,000–10,000 in stabilization basin	5–10		50–150	
Aerobic lagoons	5–10 days, longer in northern climates, less in southern climates	See Table 3.11						
Aerobic ponds	5–10 days, longer in northern climates, less in southern climates	See Table 3.11						

(Continued)

Table 3.12. (*Continued*)

	Detention time, in hours or days, as noted	BOD$_5$ loading, in lbs./day per 1,000 cf	F/M ratio, in lbs. of BOD$_5$/day per lb. of MLSS	MLSS, in mg/L	Sludge age, in days (sludge retention time)	Typical aeration period, in hours	Sludge recycle rate, without nutrient removal, in %	Typical BOD$_5$ removal efficiency, in %
Anaerobic lagoons and ponds	Minimum of 2 cells, minimum of 4 cells for flows >100,000 gal/d (400 m³/d) Minimum of 3 days per cell,	See Table 3.11						
Facultative lagoons and ponds	5–7 months in northern climates 3–6 month in southern climates Typically the primary cell is twice the size of the secondary cells.	See Table 3.11						

Oxidation ditch	3–5 hours	5–15	0.04–0.1 (MLVSS base)	3,000–5,000	15–30	75–150
Biological tower		>50 Hydraulic loadings of >1 gpm/ft^2 (>60 m/d or 0.7 L/m^2 *d)				

Source: Hammer and Hammer, (2008); Davis (2011); and Metcalf & Eddy (2014).

Table 3.11 shows the range of acceptable values for the various design parameters in a wastewater treatment plant. There are four different types of activated sludge facility listed in the table.

The following Example Problem 3.8 incorporates the calculation of most of the parameters discussed earlier, plus a few others that are useful to the proper operational control of the facility.

Example Problem 3.8

A conventional activated sludge wastewater treatment plant operates under the following conditions:

Aeration basin dimensions:	20 ft wide × 60 ft long × 8 feet deep
Wastewater flow rate:	5.0 mgd
Return activated sludge flow rate:	1.7 mgd
Waste activated sludge flow rate:	0.025 mgd
MLSS concentration:	2,500 mg/L
SS in waste sludge:	12,000 mg/L
Influent BOD_5:	150 mg/L
Effluent BOD_5:	20 mg/L
Effluent SS:	20 mg/L

Using those data, calculate the following characteristics of this facility:

1. BOD_5 loading to the plant in lbs./day
2. BOD_5 loading to the aeration basin in lbs./day/1000 cf
3. MLSS in the aeration basin in lbs.
4. BOD loading to the aeration basin in lbs./day/lbs of MLSS
5. Sludge age in days
6. Aeration period in hours
7. Return activated sludge flow rate in percent
8. BOD_5 removal efficiency for the plant, in percent
9. Mass of waste sludge generated in lbs./day
10. Mass of waste sludge produced in lbs. of sludge solids wasted/lb. of BOD_5 applied to the plant

Solution

1. BOD_5 loading to plant = 150 mg/L × 5.0 mgd × 8.34 = <u>6255 lbs/day</u>
2. BOD_5 loading to aeration basin
   V = 20 ft × 60 ft × 8 ft = <u>9600 cf</u>
   [(6255 lbs/day)/9600 cf] × 1000 = <u>651.6 lbs/1000 cf/day</u>

3. MLSS in aeration basin
   = (2500 mg/L) (9600 cf) (7.48 gal/cf) (1 mgd/1,000,000 gal) (8.34)
   = 1,497 lbs. = 1500 lbs. (approx.)
4. BOD_5 loading to aeration basin per lb. MLSS
   = (6255 lbs BOD/day)/(1497 lbs. MLSS) = 4.18 lbs. BOD/lb. MLSS
5. Sludge age = [(2500 mg/L) (0.072 million gallons of MLSS)] /
   [(20 mg/L) (5.0 Million gal.) + (12,000 mg/L) (0.025 million
   gallons)]
   = (180)/(100 + 300) = 180/400 = 0.45 days
6. Aeration period
   [(0.072 × 10^6 gal) (24 hrs./day)]/(5.0 × 10^6 gal/day) = 0.346 hrs.
7. Return activated sludge return rate in percent = [(1.7 mgd)/
   (5.0 mgd)] × 100 = 34%
8. Removal efficiency= (150 mg/L − 20 mg/L)/150 mg/L = 0.867 = 87%
9. Sludge mass produced= (12,000 mg/L) (0.025 MGD) (8.34)
   = 2,502 lbs./day
10. Sludge wasting rate =
    (2,502 lbs./day)/ (6255 lbs./day) = 0.4 lbs. sludge/lb. BOD applied

3.5.2.2 Tapered Aeration and Step Feed (Occasionally Described as Step Aeration)

As noted earlier, there are several variations possible to a conventional activated sludge treatment system. In a conventional activated sludge system, the wastewater is aerated in a circular or square basin in which the mixed liquor is constantly agitated as the air is applied. The objective is to maintain a completely mixed environment wherein the concentration of BOD_5, SS, and dissolved oxygen are constant throughout the basin. With this type of reactor, which is closely modeled as an ideal completely mixed reactor, the effluent concentration of BOD_5, SS, and dissolved oxygen are essentially equal to the concentrations at any point within the reactor. Thus, the reactor has to be large enough to maintain suitable internal concentrations for effective discharge parameters to be met at the outlet from the secondary sedimentation basin that follows.

With a rectangular aeration basin, however, things change. A rectangular basin typically does not act as a completely mixed reactor, but rather one more closely resembling a modified plug flow reactor wherein the flow comes in at one end, undergoes chemical and biological change as it moves through the reactor, and then leaves at the back end with much different concentrations than those with which it entered the reactor. Consequently, the concentrations of BOD_5 and SS concentrations are

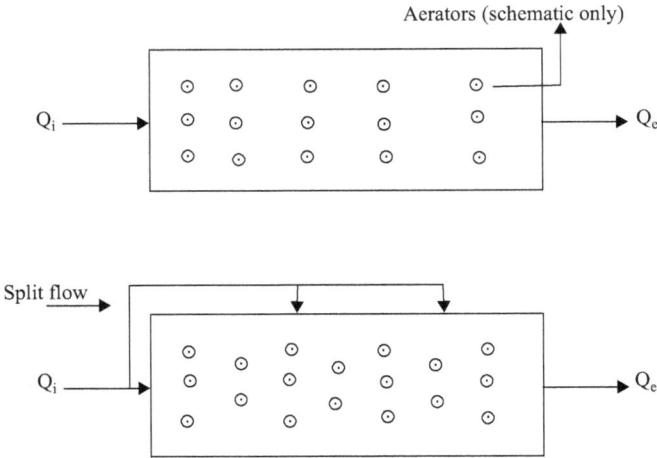

Figure 3.7. Tapered aeration (top) versus step feed (bottom) – schematic only

constantly declining as the wastewater flows through the reactor and the dissolved oxygen concentration tends to constantly increase.

The reason for using a rectangular basin is to reduce the volume of air required to maintain adequate dissolved oxygen concentration throughout the basin. The usual, or "conventional," way to do this is through a schematic called "tapered aeration." With tapered aeration, the volume of air applied is reduced as the wastewater moves through the tank—the most air is applied where the BOD_5 is the highest and less air is used where the BOD_5 is less.

A variation of that concept is based on the notion that adding influent to the aeration basin at different points along the edge of the basin, thereby keeping the BOD_5 concentration fairly constant, would allow for the use of a constant air supply, which is easier to do, and still get the same results as the tapered aeration concept. This variation is "step feed," which really means "step loading" of the reactor.

Figure 3.7 depicts these two types of aeration schematics—notice that the step feed concept splits the influent to the tank and applies it at three or four different places along the edge of the reactor

3.5.2.3 Extended Aeration

Extended aeration occurs when the wastewater is subjected to much longer aeration periods than usual, which may include excessive recycle options. The idea with this approach is to allow enough time for all, or essentially all, of the organic components of the sludge to be stabilized (biologically oxidized) inside the aeration basin. This minimizes the sludge production as a waste product, but requires as a trade-off the need to provide extensive aeration of the system. See Table 3.11 for typical design parameters. Aerated lagoons and small packaged treatment plants are typically extended aeration facilities due to the difficulty with sludge removal from lagoons and the desire to eliminate sludge handling for smaller treatment plants.

3.5.2.4 Pure Oxygen Systems

With a pure oxygen system, the aeration is done using a pure, or nearly pure, compressed oxygen feed instead of outside air. The concept is based on the notion that the microbes need the oxygen, contained in air and not the air itself since air is generally comprised of about 79 percent nitrogen and only 21 percent oxygen. Therefore, feeding pure oxygen provides a higher gas transfer rate between the aeration feed and the mixed liquor than feeding air. Therefore, a lower blower feed rate is needed and that significantly reduces aerations costs. Moreover, if the pure oxygen is supplied as a liquefied gas, the expansion of the liquid to a gas can provide sufficient pressure to force the gas into the mixed liquor, in many cases, without the need for auxiliary pumping or blowing at all.

This concept does yield a smaller footprint for the treatment plant overall since the higher resulting dissolved oxygen concentrations tend to allow for shorter aeration periods and a concurrent reduction in the size of the aeration basin. This can lead to lower capital and maintenance costs, depending upon how far away the source of the liquefied oxygen is. Liquid oxygen tends to be very expensive to ship and use, which can raise operating costs relative to those of conventional activated sludge systems.

There are some increased risks associated with the use of pure oxygen relative to the use of conventional activated sludge. Specifically, pure oxygen is toxic to humans and the environment, there are certain dangers of explosion and fire associated with handling the material, and there is a risk of supply interruption due to weather, labor strikes, and other events that could cause the supply at a treatment plant to be lost. In that event, the ability to operate the treatment plant in such a way as to achieve discharge requirements could be seriously compromised.

3.5.3 PONDS AND LAGOONS

Ponds and lagoons are used for two principal purposes: wastewater treatment, generally through an extended aeration mode, and sludge digestion.

With respect to wastewater treatment, there are five basic types of lagoon operating concepts: aerobic lagoons, aerobic ponds, anaerobic lagoons, facultative lagoons, and tertiary or polishing ponds. All wastewater treatment lagoon systems are operated with the idea of minimizing the quantity of sludge that needs to be removed because sludge removal typically requires taking the lagoon off-line, draining it, dewatering the sludge in-situ, and then excavating the dewatered sludge without damaging any aeration equipment, baffles, or the lagoon structure.

3.5.3.1 Aerobic Lagoons

Aerobic lagoons are operated with an aeration system installed in the basin to maintain a high dissolved oxygen concentration throughout the water column. These lagoons are also typically divided into two or three zones with baffles to provide what amounts to serial treatment. The effluent from the first compartment flows to the second and the effluent from the second flows to the third, and so forth. Often, several smaller lagoons are constructed in series, rather than baffling the main lagoon, to accomplish the same end result. These lagoons and ponds are typically used to cut high strength waste ahead of a facultative lagoon or secondary system or to treat industrial wastewater to a condition compatible with municipal systems. They are also commonly used in rural areas as polishing lagoons following mechanical primary treatment options.

Aeration systems may consist of plastic tubes with a series of carefully placed and evenly sized holes drilled into them that are placed across the bottom of the lagoon and anchored in place. They are often set up on small pedestals to provide space beneath them for sludge accumulation without clogging the pores. The plastic feeder lines are connected to a header along the bank. The header is connected to a blower inside a blower house and control building. Sometimes separate headers are installed on each side of the lagoon feeding intermediate tubes so that if one or more tubes clog, or there is a failure in the header for any reason, there is still an opportunity to provide good aeration. All the cross tubing is connected to both headers with shut-off valves at each end of each tube. Alternate tubes are then fed from opposite headers unless there is a problem detected. Problems are generally rather obvious because the bubble pattern on the lagoon surface will change as flow rates from the bubble tubes change.

A second method is to install aerators on the bottom of the basin, generally on risers that allow for sludge storage beneath the bubblers. These are more often rigid galvanized pipes, but may be plastic, as well. The air comes from headers, as described earlier, and is squeezed out through tiny pores in the surface of the aerator.

A debate has evolved between those who favor larger bubbles because of the reduced pressure needed to produce them and those who favor fine bubble diffusers because of the better oxygen transfer characteristics of the fine bubbles. Indeed, the fine bubble enthusiasts tend to also point out that large bubbles need to break apart into fine bubbles before any significant oxygen transfer can occur. Therefore, it is better to put the energy into creating the smaller bubbles in the first place than to push much larger quantities of air through the system at a lower pressure. Coarse bubble diffusers are typically used where a more aggressive rolling action in the reactor is desired or required for any reason.

Large bubble diffusers tend to be round caps placed on top of the riser pipe with strategically placed holes through which the air is forced by the blower. Fine bubble diffusers use a similar approach, but with much smaller openings. Sandstone bubblers, and artificial sandstone bubblers, similar to those found in a home fish tank, have also been employed successfully in lagoons.

Surface aerators are often employed in lagoons. The main issue with these devices is that they tend to keep the sludge in suspension until it is essentially oxidized to a nonorganic ash. This may be a good thing in that the sludge is more easily oxidized in suspension, but it also leads to suspended solids carry-over from the lagoon effluent. A sedimentation basin is generally required following such lagoons and it is noted that the fine ash particles seldom settle well without some form of coagulant aide.

3.5.3.2 Aerobic Ponds

Aerobic ponds are similar to aerobic lagoons except that ponds are aerated through natural actions of photosynthesis. The ponds are typically very shallow, in the order of 3 feet (1 meter) or less, and normal dissolved oxygen concentrations are maintained throughout that depth through natural biorespiration processes and natural mixing through air currents and thermal currents.

3.5.3.3 Anaerobic Lagoons

Anaerobic ponds and lagoons are not often employed because of the obnoxious odors that tend to emanate from them. They do serve as sludge digesters when they can be located sufficiently away from population centers to allow odors to dissipate adequately. Ponds and lagoons are maintained in an aerobic or anaerobic state as a function of the BOD_5 loading and mechanical aeration applied. If the BOD_5 loading exceeds the capacity of the algae to produce oxygen, reducing conditions develop that convert metal sulfides to suspended solids, and the water becomes turbid. That further reduces the algae growth and the pond goes anaerobic.

When used, anaerobic ponds are typically used to pretreat industrial wastes, particularly wastes classified as high-temperature and high strength wastes. Anaerobic processes are generally favored at higher temperatures than aerobic temperatures due to the lower dissolved oxygen content able to be maintained at higher temperatures.

3.5.3.4 Facultative Lagoons

Facultative lagoons appear to be the most common type of treatment system for rural communities with very small (typically fewer than 5,000 people) populations. They are constructed to provide very long retention times, typically in the 5 to 7 month range. The long retention times do several things. They provide time for organics reduction in the spring, following a colder period during the winter when biological activity may be significantly reduced, they provide excellent sludge reduction capacity, and they provide long-term effluent quality stability even when flows fluctuate wildly during the year.

These ponds are referred to as facultative because the BOD_5 loading is sufficiently low to allow good algal growth near the surface, which keeps the top three feet or so in an aerobic condition most of the time, while the lower 2 feet or so operate anaerobically due to a lack of mixing. Dissolved BOD_5 is best removed aerobically, but sludge decomposition works best anaerobically and the aerobic zone ameliorates noxious odors emanating from the bottom of the lagoon, thereby providing the best of both aerobic and anaerobic lagoons in one pond. It is possible, however, for noxious odors to be present during a spring thaw for several weeks until the algae are able to recover sufficient growth to provide enough oxygen to keep the water aerobic near the surface.

Approximately 3 feet of freeboard above the normal high water line is typically employed as a safeguard against overflowing in high flow periods or in high wind conditions. Facultative lagoons are also typically operated in a two or three cell series to provide proper polishing of the primary cell effluent prior to discharge.

Table 3.13. Design parameters for treatment ponds and lagoons

	Typical operating depths in ft (m)	Typical detention times in days	Typical BOD_5 loading rates in lbs./ac/d ($g/m^2/d$)
Aerobic lagoons	10–12 ft (3–3.6 m)	5–10 days longer in northern climates, less in southern climates	Based on a formula: Eff. BOD_5/Inf. $BOD_5 = 1/(1 + kt)$ where k = reaction rate constant, d^{-1} and t = detention time in days
Aerobic ponds	<3 ft (<1 m)	5–10 days, longer in northern climates, less in southern climates	Based on a formula: Eff. BOD_5/Inf. $BOD_5 = 1/(1 + kt)$ where k = reaction rate constant, d^{-1} and t = detention time in days
Anaerobic lagoons (parameter data vary widely; used mostly for commercial organic waste such as cannery or rendering wastes)	3.0–15 ft (1–5 m) typically 6–8 feet (2–2.5 m)	5–10 days for domestic waste; typically 30–60 days for organic commercial wastes	1,500–2,000 lbs. of BOD_5 per acre /d for domestic waste (2,000–2,300 kg/ha/d) 400–1,500 lbs of BOD_5 per acre/d for organic commercial wastes (440–1,650 kg/ha/d)
Facultative lagoons	Primary cells: 9–13 ft (2.7–3.9 m) in northern climates and 5–9 ft (1.5–2.7 m) in southern climates (includes sludge storage); add 5–7 ft (1.5–2.1 m) for freeboard in northern climates and 3–5 ft (0.9–1.5 m) in southern climates. Secondary cells: 5–7 ft (1.5–2.1 m) in northern climates 3–5 ft (0.9–1.5 m) in southern climates	5–7 months in northern climates; 3–6 month in southern climates; typically the primary cell is twice the size of the secondary cells	20 lbs./ac/d (2.2 $g/m^2/d$) in northern climates (to minimize odor problems during spring thaw) 50 lb./ac/d (5.6 g/m2/d) in southern climates
Tertiary lagoons	2–3 ft (0.6–0.9 m)	10–15 days	<15 lbs./ac/d (<1.7 $g/m^2/d$)

Source: Davis (2011); Hammer (2008); and Droste (1997)

3.5.3.5 Tertiary Lagoons

Tertiary ponds are generally used as polishing ponds for conventional acti-vated sludge plants or trickling filter systems. They are operated at high hydraulic loading rates (relative to other types of ponds) with detention times in the order of 10 to 15 days, and low BOD_5 loading rates and shal-low depths to allow effective oxidation without mechanical aeration.

3.6   SLUDGE MANAGEMENT

Regardless of whether a fixed film or a suspended growth biological treat-ment system is utilized, there is almost always (except with extended aeration) a significant quantity of sludge generated from the treatment of wastewater. All of that sludge needs to be managed. This section addresses the management of that sludge.

3.6.1   PRELIMINARY SLUDGE HANDLING OPERATIONS

Sludge is collected from several places in a wastewater treatment plant. Primary sludge comes from the primary clarifier, secondary sludge comes from the secondary clarifiers and any intermediate clarifiers, and tertiary sludge is generated by most tertiary systems. The sludge leaving a sedi-mentation unit is necessarily very low in solids content—typically in the 1 to 8 percent solids range (92 to 99 percent moisture content) for primary sludge, 0.5 to 1.5 percent solids content (98.5 to 99.5 percent moisture content) for settled activated sludge, and 1 to 6 percent solids content (94 to 99 percent moisture content) for trickling filter sludge. The nature of sludge from a tertiary system depends on its purpose for, and the methods used, in the tertiary process. This all means that the sludge takes up a lot of space, it is difficult to handle, and can be difficult to treat effectively. The purpose for the sludge thickener is to reduce the moisture content to a more manageable range and to provide a way to return the excess moisture back to the secondary treatment unit for further BOD_5 reductions.

It is also axiomatic that primary sludge, secondary sludge, and tertiary sludge are all very different from each other. The primary sludge is com-prised of large, heavy, settleable, or floatable solids with very high BOD_5 concentrations. This type of sludge does not degrade as quickly as other types because of the size of the particles. Conversely, secondary sludge is generally comprised of fine particles of suspended biological material and suspended solids with high BOD_5 concentrations that have not yet solubilized in the secondary system. They will degrade relatively quickly in the sludge digesters, if given a proper opportunity. Tertiary sludge may or may not be compatible with these two types and may or may not be mixed with them in the same digester.

To prepare the sludge for further treatment, the sludge from all sedimentation and skimming operations are generally mixed together in a single storage tank, typically agitated or stirred to prevent sedimentation in that unit. The storage tank is used to smooth out the flow variations in sludge generated by variable flow treatment units. The sludge is then sent to a second unit where it is subjected to grinding or maceration to break up the larger particles into a uniform size, blending to mix the primary and secondary solids together to rapidly expose the primary solids to the now hungry activated sludge microbes, and, where necessary, degritting operations. Occasionally, the storage tank is omitted if the flow of solids is consistent enough to feed a blending tank directly through a constant flow grinder pump. Necessary storage is then accomplished directly in the blending tank.

3.6.2 THICKENING

The second step in the process is thickening. Here much of the excess moisture is removed so that the sludge can be effectively digested in the smallest reasonable space. There are many systems in use to thicken sludge. Some treatment plants are designed to cothicken primary and secondary sludge. The much lighter secondary sludge does not tend to separate well by gravity, however. Hence, many plants thicken the waste activated sludge separately with gravity belt thickeners, rotary drum thickeners, or dissolved air flotation, for example, and thicken the primary sludge with gravity thickeners.

3.6.2.1 Gravity Thickeners

The simplest form of sludge thickener is a gravity thickener. The recovered biosolids are allowed to settle in the thickener and the liquid that collects at the top is recycled to the head of the secondary treatment system. The effectiveness of gravity thickening is a function of the sludge source. The best results are generally obtained from pure primary sludge because those particles are heavier and settle more easily than other particles. It is not uncommon to settle primary sludge in a gravity thickener, while using an alternative approach, such as dissolved air flotation, to thicken secondary sludge, and to then blend the two thickened flows for further treatment.

Table 3.14. Design and performance parameters for gravity thickeners

Source	Typical solids loading rate, in lbs/d-ft^2 (kg/d-m^2) of tank bottom	Typical influent suspended solids concentration, in % solids	Typical effluent suspended solids concentration, in % solids
Primary sludge (separated)	20–30 (95–150)	1–7	3–10
Conventional waste activated sludge (separated)	2.5–8 (12–40)	0.5–1.5	2–3
Trickling filter sludge (separated)	7–10 (35–50)	1–4	3–6
Rotating biological contactor sludge (separated)	7–10 (35–50)	1–3.5	2–5
Extended aeration activated sludge	5–8 (25–40)	0.2–1	2–3
Pure oxygen activated sludge	4–8 (20–40)	0.5–5	2–3
Primary sludge combined with waste activated sludge	5–17 (25–85)	0.5–4	2–7
Primary sludge combined with trickling filter sludge	10–20 (50–100)	1–6	3–9
Primary sludge combined with rotating biological contactor sludge	10–18 (50–90)	1–6	3–8

Source: Davis 2011; Hammer and Hammer (2008); and Metcalf & Eddy (2014)

That does add a level of complexity to the treatment process that may not be essential, but it can improve the performance of the thickener in many cases. Table 3.14 shows typical mass loading rates and expected results from various types of sludges and sludge blends. It is noted that waste activated sludge typically does not separately thicken well in a gravity thickener unless significant coagulation aides are used to enhance the process.

Gravity thickening should allow the solids to accumulate to a concentration of around 2 to 10 percent solids, depending on the source. Some tertiary sludges may accumulate to as much as 12 percent solids, depending on the nature of the tertiary process. It may be necessary to add a coagulant to some sludge, particularly a sludge with low alkalinity, to generate a good solids content.

3.6.2.2 Dissolved Air Flotation

It is noted earlier that pure conventional waste activated sludge does not always thicken well in a gravity thickener. This is due to the fact that the solids tend to be light and fluffy, particularly with extended aeration or pure oxygen systems. It is a truism of wastewater treatment that the easiest way to handle waste products is the same way that Nature would handle them. Therefore, if the solids want to settle in Nature, it is useful to let them settle in a gravity thickener. But if the solids want to float in Nature, then it is often more advantageous to let them float for removal in the treatment plant. That process is commonly enhanced, however, with a dissolved air flotation unit.

With this concept, the settled sludge is transferred to the bottom of a flotation tank. A pipe carrying a recycle volume of supernatant is connected to a closed air dissolution tank into which a significant volume of air is introduced under pressure. As the recycled supernatant, carrying the dissolved air from the dissolution tank, is introduced at the bottom of the flotation tank, fine air bubbles are released into the flotation tank. The fine air bubbles attach to the light sludge particles and carry them to the top of the tank in a biological foam. That foam is then skimmed off the top, typically with a moving belt scraper, the air bubbles slowly burst and the thickened solids are transferred to the next stage, often after being mixed with the gravity thickened primary solids.

Dissolved air flotation relies heavily on the design of the system for effective treatment. Various parameters of the design are important; among those of most importance are the air-to-solids ratio, the hydraulic loading rate, the polymer addition dosage, and the solids loading rate. The air-to-solids ratio is perhaps the most important of those. According to Davis, most municipal plants achieve good results with air-to-solids ratios in the range of 0.02:1 to 0.06:1 measured as a mass ratio of air available for flotation to the influent solids.

3.6.2.3 Gravity Belt Thickeners

A gravity belt thickener is comprised of a single pass, permeable belt of synthetic fabric that passes over a horizontal surface. As the belt passes

Table 3.15. Typical design and performance parameters for alternative sludge thickening options.

Thickener type	Typical solids loading rates, in lbs/d-ft² (kg/d-m²)	Typical hydraulic loading rates, in gal/min (L/min)	Typical influent solids concentration, in %	Typical effluent solids concentrations, in % (coagulation sludge)
Dissolved air flotation	120–725 without chemicals, up to 1,200 with chemicals (25–150 without chemicals, up to 240 with chemicals)	0.5–2 per ft² of surface area (20–80) per m² of surface area	0.5–1	3–6
Gravity belt filter press		100–250 (380–950) per meter of belt width	0.5–1	15–30
Rotary drum thickener		Up to 120 (up to 450)	0.5–1	4–9
Centrifuge thickener		400–600 (1,500–2,300)	0.5–4	20–25
Plate and frame press			0.5–4	30–40
Gravity thickening			0.5–1	2–4
Vacuum filter				Not commonly used for coagulation sludge

Source: Davis (2011); Hammer and Hammer (2008); and Metcalf & Eddy (2014)

over the flat surface, wet sludge is introduced across the surface of the belt, allowing free water in the sludge to drain out through the belt into a collection trough. The drainage water is returned to the influent to the secondary treatment system. The drained solids are removed from the belt with a doctor blade that scrapes the surface of the belt and they are then moved to the next treatment phase. The belt is typically backwashed with supernatant prior to return of the supernatant to the secondary treatment units.

A polymer is generally used with belt filters to assist the release of free water from the sludge. The polymers are added in a rapid mix tank just prior to discharge of the sludge to the belt surface. Fixed plows rake the sludge as the belt moves along the horizontal surface to encourage the maximum release of free water.

3.6.2.4   Rotary Drum Thickening

This type of system includes a polymer feed system and a rotating screen, covered on the inside with a synthetic liner material. Sludge is fed into the system at 0.5 to 6 percent solids, depending on the source, and exits at concentrations of 3 to 9 percent. The liquid is squeezed out through fine holes in the screen while the majority of the solids are retained inside the drum. As the drum rotates, the solids move toward the exit and fall out the back end as thickened sludge.

3.6.3   STABILIZATION

Stabilization of sludge is designed to accomplish one of two objectives. A lime stabilization process is designed to stop further bacterial degradation of the sludge and to inactivate or kill any remaining pathogenic microbes, destroy odor causing bacteria, and eliminate the potential for putrification of the solids. Aerobic and anaerobic stabilization are designed to complete the bacteriological degradation of the organic fractions. Composting is similar to aerobic digestion in its objectives. The methods use very different approaches and the chemistry is vastly different in each approach.

3.6.3.1   Lime Stabilization

Lime stabilization (also called alkaline stabilization) involves the use of quicklime or hydrated lime as an additive to the thickened sludge to increase the pH to a pH of 12 or greater and to maintain that value for 72 hours or more or until the pathogens are inactivated. The result is a nutrient-rich humus that is essentially free of pathogens. The volume of the biosolids is not reduced, and the total mass of material to be disposed is actually increased by the volume of the lime added.

The intent of sludge stabilization is to achieve the regulatory objectives of 40 CFR Part 503. To achieve a class A sludge, in addition to maintaining the pH above 12 for 72 hours, the regulations also require a temperature in the sludge of at least 52°C for a period of 12 hours while the pH is at 12 or greater. The solids must then be air-dried at elevated temperatures to no less than 50 percent solids after the 72 hour retention period is concluded. To achieve a class B sludge, the pH must be maintained at a value greater than 12 for at least 2 hours at a temperature of 25°C and greater than 11.5 for an additional 22 hours.

Generally, the sludge is treated in a liquid state then dewatered ("pretreatment"). The liquid sludge is constantly mixed during treatment using coarse bubble diffusers or mechanical mixers. Air has the potential for cooling the sludge so care must be taken to preheat the air or reduce its volume as much as possible. The lime may also be added to dewatered sludge in a pug mill. The objective of raising the pH and temperature is the same, but the dry lime is added to the dewatered solids as "post treatment."

3.6.3.2 Aerobic Digestion

The concept with aerobic digestion is to continue the conventional activated sludge biosolids degradation process to a more complete conclusion. This is done by placing the sludge into a closed tank and agitating the material with an air or high purity oxygen for a relatively long period of time. Detention times of 40 days or longer at a temperature of about 20°C and 60 days or longer at 15°C are called for by the Code of Federal Regulations. At that time, one of three tests must be met to verify that pathogen destruction and a sufficient reduction in vector attractiveness have been attained. Those tests include

a. verification that at least a 35 percent reduction in volatile solids has been achieved;
b. results of specific bench scale testing for volatile solids reduction indicates that effective treatment will have occurred;
c. a specific oxygen utilization rate less than or equal to 1.5 mg of O_2/g of total solids has been achieve at a temperature of 20°C.

Normal operations for this type of stabilization system incorporate a continuous sludge feed to the digester with intermittent supernatant and digested sludge withdrawals. Aeration is applied continuously during the filling and treatment phases, but is discontinued for a short period just prior to supernatant and digested sludge withdrawal to allow the digested solids to settle and the supernatant to clear. This is typically a daily cycle, with raw sludge added in the morning, aeration stopped for a period during the afternoon for settling, then withdrawal of supernatant for return to the head of the treatment plant, and withdrawal of digested solids, as needed. Aeration is then restarted for the overnight period. Digested solids are typically removed when the tank does not gravity settle well and does not yield a clear enough supernatant. At that time a portion of the settled solids is removed for disposal.

3.6.3.3 Anaerobic Digestion

Anaerobic digestion is a more complex process than aerobic digestion. There are, in fact, three distinct phases to anaerobic digestion. The first phase (the *acidogenesis* or *acetogenesis* phase) involves the hydrolyzing of complex waste components, such as fats, proteins, and polysaccharides, into component subunits, such as fatty acids, amino acids, triglycerides, and sugars. The byproducts of hydrolysis are then converted through a fermentation process into simple organic compounds, such as short-chain acids and alcohols. In the second phase (the *acid fermentation* phase), the organic compounds created through the initial fermentation stage of the first phase are metabolized through further fermentation into organic acids, alcohols, and new bacteria cells. In the third phase (the *methane fermentation* phase), the end products of the second stage fermentation are converted to methane, carbon dioxide, and minor quantities of other miscellaneous gases.

With an anaerobic process, the biological degradation is carried out in the absence of significant supplies of oxygen. Most anaerobic digesters, but not all, are operated as suspended growth reactors operated as completely mixed reactors.

There are two temperature ranges typically encountered in municipal sludge management. Higher temperature, thermophilic operations, typically in the range of 120°F to 135°F (50°C to 57°C), require smaller reactor sizes due to the increased biological activity at the higher temperatures. They also do a better job of destroying pathogens and reducing the volume of residual solids. Higher temperature digestion is not widely practiced, however, due to the significant increase in energy costs associated with the maintenance of the higher temperatures.

Somewhat lower temperature operations, in the mesophilic range, typically in the range of 85°F to 100°F (30°C to 38°C), take longer than thermophilic digestion, and therefore also require a larger digester, but can usually accomplish the same regulatory end product results at a lower overall energy cost.

One of the biggest benefits of anaerobic digestion is related to its ability to produce energy in the form of methane gas. The gas is typically burned to generate heat to keep the digester temperature up, and, where cost-effective, can be used to cogenerate electricity and heat for in-plant use (a space heating loop, for example) as well. Anaerobic digestion typically requires gas storage, digester heating, and digester mixing systems. Mixing can be via mechanical mixers, pumps, gas nozzles, or large gas bubble "cannons." Therefore, anaerobic digestion can be considered a "green" technology and can be combined with loading of other locally available organic wastes (such as fats, oils and greases, and food wastes) for additional energy recovery.

3.6.3.4 Temperature-Phased Digestion

A more recent German process, typically referred to as *temperature-phased digestion*, utilizes both thermophilic and mesophilic digestion in a combined two-stage digestion process. The reactors are typically sized to operate either in a mesophilic-thermophilic mode, or in a thermophilic-mesophilic mode. The concept is to take advantage of the rapid destruction of organics and pathogens in the thermophilic range, while reducing the overall energy costs through a slower, but smaller, mesophilic reactor. This type of reactor arrangement has been shown to absorb shock loading better than single temperature reactors and to produce more consistent end products, all of which meet regulatory requirements for class A sludge.

3.6.4 CONDITIONING

Following digestion, the sludge needs to be further dewatered to reduce the cost of handling and disposal, which is typically a chemical process in which one or more polymers are added to the liquid sludge. The function of the polymer is to break the bond between the organic and inorganic solids in the sludge and the water molecules that bind them together. The water is then able to properly separate from the solids in the following dewatering phase. Chemical addition is done in a mixing reactor following digestion, or through an in-line mixer as the sludge is moved from the digester to the dewatering equipment.

3.6.5 DEWATERING AND FINAL DISPOSAL

Sludge is further dewatered prior to final disposal. The purpose of dewatering here is to reduce the volume to be handled as trucked waste. It is important to recognize that the trucked sludge does have to flow into and out of the hauling vehicle, however, and that too much dewatering is not a good thing. Typically a minimum of about six percent solids is considered the limit for trucking free-flowing sludge. Drier material would require mechanical removal from the truck or trailer. Dewatering is typically done using one or more of various types of equipment, such as belt presses, plate and frame presses, centrifuges, or sludge drying beds similar to those described in Table 3.15.

Final disposition typically involves: soil addition to loosen tight soils so that crops can be grown effectively, reclamation of inorganic soils at sites such as strip mining operations where there is no top soil otherwise remaining, land spreading to restore inorganic nutrients to soils and assist in retaining soil moisture during a growing season, incineration, pelletizing for easier soil application, or landfilling. Land disposal is controlled by state regulations and every state has its own requirements. Reference to specific state regulations will be necessary in each case.

In more rural areas, composting of sludge is done with wood chips or other bulking agent to produce a product that can be land-applied. In more urban areas, dewatered sludge is often incinerated. Where incineration is employed, it is advantageous to maintain as much heat value in the sludge as possible (undigested sludge burns more readily than digested sludge). Still other communities heat-dry digested sludge to produce a product that can be applied to the soil.

3.7   TERTIARY TREATMENT UNITS

Tertiary treatment of domestic wastewater typically involves the reduction of specific nutrient concentrations for the protection of disposal sites or receiving waters. The tertiary treatment of industrial wastewater can involve the same principles, but may also require the reduction or removal of toxins not generally found in domestic wastewater.

3.7.1   INTERMEDIARY SEDIMENTATION BASINS

The reduction of suspended solids beyond that normally required for surface water discharge of treated sewage is done with an intermediate or polishing clarifier. Intermediate clarifiers and polishing clarifiers are designed the same way as a secondary clarifier using the design parameters found in Table 4.1 (Chapter 4).

3.7.2 NUTRIENT REMOVAL TECHNIQUES

Nutrient removal generally implies the removal of nitrogen or phosphorous. Nitrogen and phosphorus in wastewater effluent can have a fertilizing impact on the receiving water. This can lead to excessive growth of algae in the receiving water. In fresh water impoundments, this can result in green "pea soup" conditions. Also, as algae die off and sink to the bottom of the impoundment, cell decomposition can lead to hypoxia (lack of dissolved oxygen) in the lower levels of the water, which is extremely deleterious to the aquatic environment (this has been found to be the case in Long Island Sound, off the coast of Connecticut and New York City, for example). As a general rule, it has been found that phosphorus is the limiting nutrient leading to eutrophication in freshwater aquatic systems and nitrogen is the limiting nutrient in marine (salt water) environments. Therefore, depending on the receiving water, there may be a need to meet low effluent requirements for N, P, or both (e.g., discharge to freshwater river impoundments that ultimately discharge to tidal marine waters).

Phosphorous has become more of an issue in recent years due to non-point discharges to rivers and streams causing significant algal blooms. Phosphorous control initially came about as a regulatory concern because nitrogen discharge control did not reduce the incidence of algal blooms in all receiving waters to the degree expected. Further investigations indicated a phosphorous limited growth phenomenon rather than, or in addition to, a nitrogen limited growth phenomenon. Consequently, phosphorous has been subjected to much greater scrutiny of late, and processes to reduce phosphorous to previously unattainable concentrations are being developed, and improved regularly.

Phosphorus is removed either chemically or biologically. Both processes involve changing a solubilized or liquid form of phosphorus (generally an ortho-phosphate) to a solid form. Chemically this is most often accomplished using either an iron salt (such as ferric chloride or ferrous sulfate) or an aluminum salt (such as alum, sodium aluminate, or poly-aluminum chloride). Lime can also be used to precipitate phosphorous through an increase in the pH in the water. This does not generally reduce the concentrations low enough to meet regulatory standards, however, and then requires an additional recarbonation step to lower the pH. Any chemical process also requires chemical storage facilities, chemical feed pumps, and extra equipment maintenance, as well as larger or additional sludge removal and processing equipment.

Biologically, phosphorus is removed from solution and concentrated inside bacterial cells. Treatment involves two different processes. The first is a fermentation zone where short chain organic compounds are created, notably volatile fatty acids (VFAs) that are attractive energy sources for phosphate accumulating organisms (PAOs). This zone also provides the opportunity for the uptake of those volatile fatty acids by the single celled PAOs. The second step requires an oxygen-rich zone where the PAOs oxidize the volatile fatty acids that they ingested in the fermentation zone as an energy source, thereby removing the soluble phosphorus from the solution.

TSS control is important during phosphorous removal because both chemical and biological phosphorus removal results in TSS containing up to 5 percent total-phosphorus. Achieving a total-P limit of 0.2 mg/L, which assumes a residual, post-treatment, soluble phosphorus concentration of 0.05 mg/L, requires an effluent TSS concentration of no more than 3 mg/L.

Nitrogen is almost universally removed biologically. The process results in the conversion of solubilized (liquid) organic nitrogen (such as urea, which is the most common form of organic nitrogen) to nitrogen gas. The nitrogen escapes harmlessly to the atmosphere; ambient air is 75 to 80 percent nitrogen gas.

In some unusual cases, the organic nitrogen readily converts to ammonium during normal treatment at the treatment plant. Those wastewater treatment plants convert ammonia to nitrate (with nitrite as an intermediate product) in highly aerobic conditions such as those created by two stage trickling filters or extended aeration treatment plants. The nitrate is then converted to nitrogen gas under anoxic (low oxygen) conditions. Unlike phosphorus removal, nitrogen removal does not result in a sludge that must be removed since the nitrogen escapes into the atmosphere as a gas.

3.8 DETAILS OF DISINFECTION OF WASTEWATER

The disinfection of wastewater, regardless of its future intended use or reuse, is an art that has evolved over many decades. Solar disinfection, in essence exposure to sunburn, was used in ancient times for this purpose. In developing areas of the world, it is not uncommon to use solar disinfection for drinking water. This technique is effective, but takes an inordinate amount of time for most wastewater flows, and the area needed to effectively expose the volumes of wastewater to solar radiation would be enormous. Consequently, alternative means have been developed for this operation.

3.8.1 ROLE, GENERATION, AND USE OF CHLORINE

Chlorine is, perhaps, the most widely accepted disinfectant in use today. The general public understands that chlorine is effective, people use it in their home swimming pools, people know it is used to treat drinking water supplies, and they believe it is safe. Chlorine is relatively inexpensive, and compared to other options it is generally accepted by the public, it is easily obtained, and it does work against most bacteria. It is not necessarily as effective against many viruses.

Chlorine is generally delivered to a wastewater treatment plant in 20-ton cylinders of liquid chlorine. The normal state for chlorine is a gas; therefore, liquefying it requires significant pressures. The gas cylinders delivered to treatment plants hold those pressures while the gas is released as a liquid through a regulator to an injection port where it is introduced directly into the waste stream or into a side stream that rapidly mixes with the main waste stream.

Issues with the use of chlorine include the fact that it is under great pressure and in the event of a malfunction in the regulator, or accidental damage to the valving at the outlet of the tank, huge quantities of chlorine gas can be very rapidly released to the environment surrounding the tank. Chlorine gas is highly lethal in concentrations well below what could be expected from a sudden release. Consequently, regulations now require that chlorine storage be inside a separate sealed room that has an access port only to the exterior of the building. All penetrations of the walls of that room need to be sealed against a pressurized release and a pressure relief system needs to be installed to release pressure in the event of a malfunction. A full face, supplied air respirator needs to be stored near the entrance to the room so that in the event an operator needs to enter the room during a leak emergency, the operator will have adequate breathing protection.

There are certain other issues associated with the use of chlorine. The residual chlorine has the potential to react with environmental factors to create trihalomethanes in the receiving water. These can be extremely harmful to aquatic life and are carcinogenic if ingested by humans. In addition, chlorine is highly corrosive and its use leads to significant maintenance issues at the treatment plants. Because of these down-sides, some communities only require seasonal disinfection with chlorine (during the summer fishing/swimming/water-contact season). In other cases, a dechlorinating agent such as sulfur dioxide or sodium bisulfite is added after the required chlorine contact time has been achieved, to react with and remove any remaining free chlorine.

Many facilities have been switching over to sodium hypochlorite solution to reduce the risks associated with gaseous chlorine. This is generally delivered in bulk to a storage tank on site, or it is generated at the point of use on the site. Since hypochlorite degrades over time, the stock must be used up and rotated in a reasonable time frame. In order to get adequate disinfection with chlorine (including hypochlorite), it is critical to demonstrate sufficient detention time at the required concentration dose. This can be achieved in a detention basin, designed to reduce short-circuiting, in the outfall pipe, or both.

3.8.2 ROLE, GENERATION, AND USE OF OZONE

An alternative to chlorine that evolved in the late twentieth century is the use of ozone as a disinfectant. Ozone has the advantage of breaking down rather quickly in the environment into oxygen molecules. That is a good thing because the energy to go from three oxygen atoms to two comes from oxidizing organic matter in the water, such as bacterial pathogens and other organics that could otherwise create an oxygen sag in the receiving water.

Ozone does not store well, so it is generated at the point of use. That allows for much higher use efficiency, but much high generation risks. Ozone is generated by passing dried air through an electrical corona. The corona is created by passing a very high voltage electrical current through a thin wire at a very low amperage. The energy to create three atom molecules from two atom molecules comes from the corona. If the air is insufficiently dried prior to passing through the corona, or if the electrical wires forming the corona get too close to each other due to vibrations, a spark-over can occur between the wires, causing a devastating explosion. Proper automatic monitoring of the air entering the generator can minimize this risk, but it does still exist.

In addition, the generated ozone is generally bubbled into the wastewater stream through stainless steel nozzles and bubblers. As it rises through the water column, the ozone disintegrates and destroys the target organisms. Not all the ozone is necessarily destroyed in the process, however, and free ozone is toxic to people. Consequently, it is necessary to collect all excess gas that passes through the wastewater and to ensure effective destruction of the residual ozone prior to release of the off-gasses to the atmosphere. Davis (2011) reported that destruction is generally done thermally with a catalyst. With a catalyst, destruction at temperatures around 48°F to 71°F (30°C to 70°C) is effective.

There is no residual with the use of ozone; therefore, many regulatory agencies required secondary chlorination anyway to ensure some small residual disinfectant in the effluent discharge.

3.8.3   ROLE, GENERATION, AND USE OF ULTRAVIOLET LIGHT

Ultraviolet light disinfection of drinking water has been practiced since the mid-twentieth century. UV was originally used in places such as fish hatcheries, where very high destruction rates of a multitude of pathogens is required as part of the ultra-purification systems employed at such facilities. It is also used extensively in drinking water disinfection. It has been generally accepted that the suspended solids concentrations in secondary wastewater treatment plant effluent are too high for effective light penetration and that the lack of a residual with this method means that there is no convenient way to verify the long-term effectiveness of the treatment.

Better filtration methods and lower suspended solids concentrations have allowed some plants, particularly smaller, remotely located facilities, to demonstrate effective use of this technology and it is gaining favor since the power to operate the system can sometimes be generated from solar panels and the "green" nature of the combined technologies is selling well with the general public. The main advantages of UV (compared to chlorine) are that it eliminates safety issues related to handling chlorine and does not require a lengthy contact time for a chemical reaction. This technology requires a lot of power available around the clock, which is why solar power is not used more extensively or for larger facilities. In general, UV is more expensive than chlorination, but it can become more cost competitive at facilities where dechlorination would also be required. Turbidity of the wastewater effluent continues to be a major factor in determining the energy life cost, despite the advances in effluent suspended solids reduction. UV can be accomplished either with a closed-vessel system (often used for smaller plants with lower flows) or in open channels with a lamp array.

3.8.4   RADIATION

Radiation has been practiced at water treatment plants for many years and it has been used to disinfect sludge at various wastewater treatment facilities over the years. It can be dangerous to generate radiation, and there does need to be a protective barrier between the source and any personnel working in the vicinity. Just like an excess of sunlight or x-rays can cause cancer, so can an excess of radiation used to disinfect wastewater. The rays will penetrate suspended solids if gamma rays or neutron rays are used. This technique is not widely used since the generators are expensive and the radiation can pose a significant risk to operators.

BIBLIOGRAPHY

Davis, M.L. 2011. *Water and Wastewater Engineering: Design Principles and Practice.* New York, NY: McGraw Hill Book Co.

Droste, R.E. 1997. *Theory and Practice of Water and Wastewater Treatment.* New York, NY: John Wiley & Sons, Inc.

Felder, R.M. 2000. *Elementary Principles of Chemical Processes.* 3rd ed. New York, NY: John Wiley & Sons.

Greywater Action for a Sustainable Water Culture. 2014. Retrieved from greywateraction.org: http://greywateraction.org/content/about-greywater-reuse (April 21, 2014).

Hammer, M.J. 2008. *Water and Wastewater Technology.* Upper Saddle River, NJ: Pearson Prentice Hall.

Managers, W.C. 2004. *Recommended Standards for Wastewater Facilities.* Albany, NY: Health Research, Inc., Health Education Services Division.

McGhee, T. 1991. *Water Supply and Sewerage.* 6th ed. New York, NY: McGraw-Hill Book Company.

Metcalf & Eddy. 2003. *Wastewater Treatment Engineering: Treatment and Reuse.* 4th ed. New York, NY: McGraw-Hill Publishers.

Metcalf & Eddy/AECOM. 2014. *Wastewater Engineering Treatment and Resource Recovery.* New York, NY: McGraw-Hill Publishers.

Rice, E.W. 2012. *Standard Methods for the Examination of Water and Wastewater.* Washington, DC: American Water Works Association/American Public Works Association/Water Environment Federation.

Richardson, S.D. 2003. "Disinfection By-Products and Other Emerging Contaminants in Drinking Water." *TrAC Trends in Analytical Chemistry* 22, no. 10, pp. 666–684. doi: http://dx.doi.org/10.1016/s0165-9936(03)01003-3

Water Environment Federation. 1998. *Design of Municipal Wastewater Treatment Plants.* Manual of Practice No. 8, 4th ed. Alexandria, VA: Water Environment Federation.

CHAPTER 4

SEDIMENTATION FUNDAMENTALS

4.1 INTRODUCTION

Wastewater treatment is designed to remove most of the organic BOD_5 (5-day biological oxygen demand) and essentially all of the inorganic constituents that enter the plant in the form of sewage. That is generally done by allowing the inorganic fractions, which are generally heavier than water, to settle out in the grit removal system and the primary clarifier. The organic fractions then move on to the secondary system where they are converted to biomass. The biomass, or sludge, is removed by settling in intermediary and secondary sedimentation basins.

Sedimentation takes on two distinctly different forms in wastewater treatment. In the primary clarifier and grit removal system, the grit and inorganics settle out using "discrete particle sedimentation." During the conversion of the organic fractions to biomass, however, those organic fractions form significantly lighter particles that settle very differently from the discrete particles. This is called "flocculent settling." This chapter describes some of the fundamental theory behind both kinds of settling.

4.2 DISCRETE PARTICLE SEDIMENTATION

Sedimentation, or clarification, is the removal of suspended particulate matter, chemical floc, and precipitates from suspension through gravity settling. Settling basins are designed on the basis of detention time, overflow rate (which is the flow into the reactor divided by the surface area of the reactor), weir loading (which is the flow out of the reactor divided by the length of the weir), and (with horizontal flow tanks, but not vertical flow tanks) the horizontal velocity.

Since flow is generally shown on a daily flow basis, but design criteria for settling basins are in hours, the detention time, calculated as V/Q (where V = volume in cubic feet and Q = flow in million gallons per day [mgd]), must be multiplied by 24 hours per day (or the Q must be divided by 24 hours per day) to get an average hourly flow rate; units must be consistent throughout.

Surface overflow rates (SOR) are calculated by dividing the Q by the surface area of the reactor. This yields units of cubic feet per square foot per day (or ft/day) or cubic meters per square meter per day (or m/day). That means that the units of SOR are units of velocity. That turns out to be very important in the design of the basins because only those particles with vertical settling velocities greater than the overflow rate, expressed as a velocity, will settle out; most of the lighter ones will go over the effluent weir and not be captured.

4.3   FLOCCULANT PARTICLE SEDIMENTATION

Flocculation is the culmination of a chemical reaction between dissolved particles and a coagulant added to the treatment unit. The flocculant changes the electrical charge on the dissolved particle, or changes the pH of the water sufficiently for the dissolved component to aggregate into a sufficiently large agglomeration of atoms to form a (usually) visible suspended solid. As particles increase in size and begin to settle, they bump into smaller particles, adding the smaller mass to the growing suspended particle and fall slightly faster. This step is often referred to as coagulation and the two terms, flocculation and coagulation, are also often used together to describe the overall phenomenon occurring.

Flocculated particles tend to be thin plate-like particles that settle very slowly. Their surface area to mass ratio tends to be very high relative to discrete particles. Consequently, the settling of flocculant particles has been described as something akin to corn flakes settling through water. As a result, the design of a flocculant particle settling basin is quite different from that for discrete particles. Both design processes are further described later in detail.

4.4   WEIRS

Weir loading is computed by dividing the average daily flow by the total weir length in feet or meters. This yields gallons per foot per day or cubic meters per meter per day as the units. Most weirs in use today are v-notch weirs. These allow for water to exit over the weir, while retaining some floating materials that pass over or under a scum baffle behind the weir.

(The v-notch weir should not be considered a substitute for removal of floating material, however.) It is not uncommon to find filamentous algae growing on the teeth of the weir in long strands. It is good practice to remove these as they develop since even though they may be adding to the biological treatment of the water passing through them, they also tend to break off regularly and exit the reactor as large clumps.

Weirs can also be straight edge weirs that act much like a dam spill-way. The advantage of that arrangement is that there is no convenient place for filamentous algae to grow, but they also do not hold back other floatables that might otherwise be retained by a v-notch weir. Since all weirs should be isolated by a floatables dam, anyway, this is often not a major problem.

4.5 CLARIFIER DESIGN

Figure 3.2 shows schematics of both a circular clarifier and a rectangular clarifier. Figure 4.4 shows the details of an upflow circular clarifier. Note that sludge is collected from the bottom of all three types of clarifier and that there is a moving arm that travels along the bottom of the rectangular clari-fier or around the bottom of the circular clarifier to nudge the soupy sludge into the end hopper from which it is removed by gravity flow or pumping.

The United States Environmental Protection Agency (EPA), Great Lakes Upper Mississippi River Board (GLUMRB), and various authors have established standards for settling basin design. Table 4.1 describes the key factors involved in the design and operation of various types of clarifiers and sedimentation basins. Redundant units capable of indepen-dent operation are generally required for all plants where average design flows exceed 100,000 gpd (380 m³/day)

Example Problem 4.1 describes how these parameters interrelate and how to check their compliance with GLUMRB standards. As usual, mgd stands for million gallons per day and gpm stands for gallons per minute.

Example Problem 4.1

Given a treatment plant with the following characteristics, calculate the major parameters used in sizing the primary clarifier and compare the results to GLUMRB standards.

  Flow = 6.0 mgd (average)
  Flow = 8.0 mgd (peak)

Table 4.1. Typical design characteristics of sedimentation basins and clarifiers

Clarifier type	Design basis	Detention time in hours	Horizontal velocity in ft/min (m/min)	Weir loading gal/lf/ day(m³/m/d)	Typical surface overflow rate gal/day/sf (m³/d/m²)	Typical side wall water depths in ft (m)
Primary clarifier	Discrete particle	1.5–2.5 (maximum settling has been observed at about 0.5 hr of detention in many cases)	4–5 (1.2–1.5)	10,000–40,000 (125–500)In practice rates of 9,600–15,300 (120–190) have been reported(GLUMRB = 10,000 gal/lf/d)	785–1,200 (dry weather without sludge recycle to primary) (32–49)1,200–3,000 (peak flow) (49–122) 600–800 (dry weather with sludge recycle to primary) (24–32)1,000–1,800 (peak flow) (40–70) (GLUMRB = 1,000 gal/d/sf at average flow and 1,500 at peak flow)	7–15(2–5 m) (GLUMRB = 7 ft[2 m]])
Intermediate clarifier	Flocculant			GLUMRB = 10,000 gal/lf/d for plants smaller than 1 mgd and 20,000 gal/lf/d for plants greater than 1 mgd	GLUMRB = 1,000 gal/day/sf(40 m³/d/m²)	GLUMRB = 7 ft(2m)

Process	Particle type				Overflow rate	Detention time
Secondary or final clarifier(general)	Flocculant particle	>4	<0.5 ft/min(<0.15 m/min)	<20,000(<75)	400–800 (dry weather)(16–33) 980–1,700 (peak flow)(40–70)	9.5–18(3–5.5)
Following conventional activated sludge	Flocculant particle				400–700(16–29)	13–18(4–5.5 m)
Following extended aeration	Flocculant particle				195–400 (dry weather)(8–16) 600–800 (peak flow)(24–32)	13–18(4–5.5 m)
Following an oxidation ditch	Flocculant particle				300–400 (average flow rate)(12–16)	13–18(4–5.5)
Following a trickling filter or RBC	Flocculant Particle				235–800 (average flow rate)(9.5–32) higher rates generally apply to deeper sidewall depths	7–10(2–3.1) for SOR <1,650 gpd/sf (6.7 m³/d/m²) 11–16(3.1–4.6) for SOR >1,650 gpd/sf (6.7 m³/d/m²)
Tubular and lamella clarifiers	Discrete particle				360–720(15–30)	

Source: Droste (1997); Davis (2011); Metcalf & Eddy (2014); and Hammer and Hammer (2004)

Clarified diameter = 90 ft
Side wall depth = 7 ft
Single weir on tank periphery

Solution

Calculate the SOR as follows:

Surface area of clarifier = π d^2/4 = 6,362 sf
Clarifier volume = 6,362 ft × 7 ft = 44,532 cf = 0.333 × 10^6 gal
SOR = 6.0 × 10^6 gal/d/6,362 sf = 943 gal/d/sf at average daily flow
SOR = 8.0 × 10^6 gal/d/6,362 sf = 1,257 gal/d/sf at peak daily flow rate

These rates are close to, but less than, the recommended maximum SOR for average flow rate established by GLUMRB and within the range for most other reference sources. At peak flow rate, they are also below the recommended maximum flow rates for both GLUMRB and most other references.

Calculate the hydraulic detention time as follows:

t_d = V/Q = 0.333 × 10^6 gal/6 × 10^6 gal = 0.056 days = 1.3 hour at average flow rate
t_d = V/Q = 0.333 × 10^6 gal/8 × 10^6 gal = 0.042 days = 1.0 hour at peak flow rate

Detention time is generally not used as a design parameter because it is defined by overflow rates and volume. These detention times should yield excellent results, however, based on the observed maximum settling at approximately 0.5 hour in most cases.

Calculate weir overflow rate as follows:

Weir length = Circumference of the tank since the weir typically hangs along the outside of the tank perimeter
Weir length = 2 π r = (2) (π) (45) = 283 feet
Weir loading = 6,000,000 gal/d/283 ft = 21,200 gal/lf/d at average flow
Weir loading = 8,000,000 gal/d/283 ft = 33,600 gal/lf/d at peak flow rates

These rates are greater than the maximum overflow rates recommended by GLUMRB, but well within the range recommended by most other references.

4.6 FLOCCULATOR CLARIFIERS

A more recent development has been a design that tries to incorporate both the flocculation phase in the shroud and the settling phase in the upflow section of a single circular sedimentation basin. This design concept is more typical where circular tanks are used, but the design has to account for the smaller settling volume than the tank would suggest due to the detention time required for the flocculation to occur inside the shroud. The shroud also has to be large enough to allow the required flocculation to occur. A rapid mix and flocculation basin concept incorporated into the shroud or at the influent end of a rectangular clarifier can help to overcome these issues. These reactors are most commonly found in water treatment facilities, rather than in wastewater treatment facilities, because the floc particles tend to be different in the two types of treatment plants and the settling characteristics of the water treatment floc are more amenable, in general, to a combined treatment unit.

4.7 DESIGN OF DISCRETE PARTICLE CLARIFIERS

Discrete particles include sand and grit, large coffee grinds, and other relatively large solid particles in the wastewater. Many of the inorganic particles settle out in the grit removal system, but smaller discrete particles take too long to settle in the short duration grit chamber. Moreover, those particles that do settle in the grit chamber tend to be inorganic, while the smaller particles tend to be organic in nature. It is useful, therefore, to try to ensure that they settle in different locations so that they can be managed successfully more easily.

There are several factors at work in any sedimentation basin. They include gravity pulling downward on the particles, friction between the settling particles and the water, a horizontal velocity factor associated with the inflow velocity, and a vertical velocity component associated with the gravitational pull. In addition, wind and thermal currents can be strong at various times of the day or year, ionic charges on the suspended particles can be problematic, and the viscosity of the fluid, as a function of temperature, typically, can also affect sedimentation significantly.

The first component examined here is the "terminal velocity" of the settling particles. Isaac Newton found the terminal velocity of particles follows the equation:

$$v = [4g(\rho_s - \rho)\, d/3C_d\rho]^{1/2} \qquad (4.1)$$

Where:

v = Terminal settling velocity
g = Gravitational constant (32.2 ft/sec² or 9.81 m/sec²)
ρ_s = Mass density of the particle
ρ = Mass density of the fluid
d = Diameter of particle
C_d = Drag coefficient

The drag coefficient is defined as follows:

$$C_d = (24/N_R) + (3/\sqrt{N_R}) + 0.34 \qquad (4.2)$$

Where:

N_R = Reynolds number

and

$N_R = vd\rho/\mu$

Where:

μ = Absolute velocity of the fluid.
The other terms are as defined as before

This equation for the Reynolds number holds up to a value of about 1,000; but if the value of N_R is less than about 0.5, the last two terms of C_d can be neglected. By neglecting those terms and manipulating the equation a bit, it is possible to get to the following simplified equation for the terminal velocity:

$$v = [(g/18\mu)\,(\rho_s - \rho)\,d^2] \qquad (4.3)$$

This is "Stokes law." All of this assumes that the particles are all round. Most wastewater particles are not round. Therefore, these equations do not apply directly. Fortunately, this is not a big problem and they can be used anyway since the variation in the answer is sufficiently miniscule as to be irrelevant in the context of wastewater treatment. The small particles that wastewater treatment is designed to capture do follow the rules and the larger ones settle so much better that the dynamics are not important and they will all be captured.

The actual design process is based on the concepts of an ideal rectangular settling basin. It assumes that the incoming particle has a horizontal velocity equal to the horizontal velocity of the incoming fluid, described by the following equation:

$$v = q/a = q/w \times h \tag{4.4}$$

Where:

v = Horizontal velocity
q = Flow rate into the basin
a = Vertical cross-sectional area of the basin defined by the width times the depth

If a particle is to be removed in the basin, the resultant of the horizontal velocity and the terminal settling velocity must be such that it will reach the bottom of the settling zone before reaching the end of the zone (see Figure 4.1).

The slope of the velocity vector in this diagram is given by the following equation:

$$v_o/v = h/l \tag{4.5}$$

Where:

v_o = Terminal velocity of each particle
v = Horizontal velocity of each particle
h = Depth of the settling zone
l = Length of the settling zone

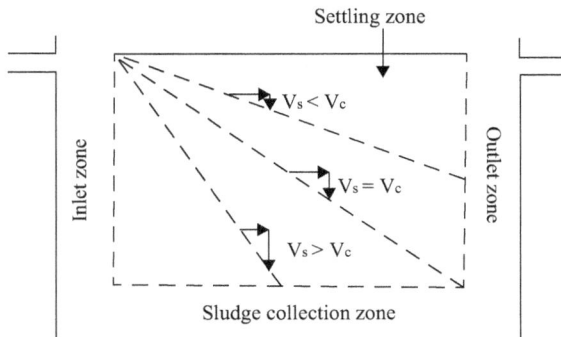

Figure 4.1. Classic representation of the influent zone, settling zone, and effluent zone of a rectangular settling basin.

This equation, by appropriate manipulation and mathematical magic, can then be rewritten as follows:

$$v_o = vh/l = (h/l)(Q/wh) = Q/wl \qquad (4.6)$$

Where:

v_o, v, h, and l are as defined previously
w = Width of the basin
Q = Flow rate into the basin

This equation is also the expression for the incoming horizontal velocity, or the

"surface overflow rate" or SOR.

That expression encompasses the terminal settling velocity, the flow into the reactor, and the surface area of the reactor. Hence,

The SOR is the settling velocity of the slowest settling particle that will be captured with 100 percent efficiency.

It is further noted that nothing is actually overflowing in the basin, but rather the incoming wastewater is being evenly distributed across the surface of the reactor.

What this expression says with respect to particles with settling velocity less than the overflow rate is that some will be captured, but not all. That is because the particles enter the settling zone in Figure 4.1 across the entire cross-sectional area of the basin. Those that enter halfway down the face of the cross-section have only half as far to settle before being captured as do the smallest target particles (those for which 100 percent capture is desired). Thus, those that enter halfway down the face need to settle only half as fast as the target particles and they will still be captured. That is similarly true for particles that enter near the bottom of the cross-sectional area; they have a very short distance to settle before being captured, but a long horizontal distance in which to do that, so they can settle at a much slower rate and still be captured. Thus, a certain percentage of those particles with settling velocities less than the overflow rate will also be captured. That percentage can be determined from a graph similar to the one in Figure 4.2.

This diagram is a plot of the terminal velocities of the various sized particles in the mix. The percentages shown are cumulative percentages of

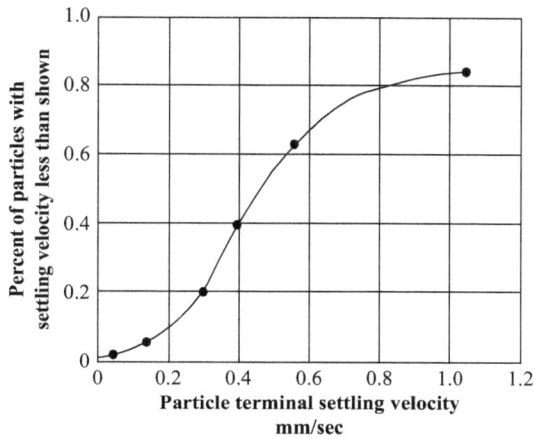

Figure 4.2. Representative plot of the terminal velocities of various sized particles in a wastewater particles.

the particles in the total particle mix that have a settling velocity equal to or less than the velocity of the particle size being plotted.

For example, from the diagram:

40 percent of particles have a settling velocity of 0.4 mm/sec or less.
85 percent have a settling velocity of 1.1 mm or less.

The 85 percent group includes the 30 percent group, plus an additional 55 percent of the total particles.

To determine the total percentage of particles captured by this basin, the percentages are plotted versus settling velocity for each selected particle size. That area above the curve is then integrated from the zero intercept to the point on the curve represented by the design overflow rate or design terminal velocity—the terminal velocity of the smallest target particle for which 100 percent capture is desired.

The equation for the total fraction captured then becomes the following:

$$f = (1 - X_s) + (1/v_s)\Sigma(v\,\Delta x) \tag{4.7}$$

Where:

f = Fraction captured
X_s = Fraction with settling velocity equal to or greater than the design SOR
v_s = Design terminal velocity
v = Terminal velocity at any point on the graph
Δx = Average fraction of particles with a settling velocity of v.

These concepts are best illustrated by the following example problems. The information provided is the grain size distribution with the particle grain size in millimeters along the top and the cumulative weight percentage of grains larger than the stated size. Thus, it can be seen that 10 percent of the particles are larger than 0.14 mm and that 95 percent are larger than 0.03 mm.

Example Problem 4.2 for Discrete Particle Settling

A settling basin is designed with a SOR of 28 m/day. Determine the overall efficiency of this clarifier if the influent flow stream has the following characteristics. Assume that Stoke's law applies and that the Reynold's Number of the system is not an issue.

Grain size distribution per the following chart
Particle specific gravity of 1.20
Water density of 0.98
Water viscosity of 1.01

Particle size in mm	0.14	0.12	0.08	0.06	0.05	0.03	0.01
Weight % > than	10	15	20	35	60	95	100

Solution

Calculate the maximum settling velocity as a function of the SOR as follows:

$$28 \text{ m/d} \times 1 \text{ d/86,400 sec} \times 1,000 \text{ mm/m} = 0.32 \text{ mm /sec} = v_s$$

Calculate the settling velocity of each particle as a function of its diameter from Equation 4.3 as follows:

$$v = (g/18 \ \mu) \ (\rho_s - \rho) \ d^2$$

Where:

$$\rho_s = (1.20) \ (0.98) = 1.18$$

Then:

$$v = [9810/18 \ (1.01)] \ (1.18 - 0.98) \ d^2$$
$$v = 107.92 \ d^2$$

Re-set the table as follows:

Particle size, in mm	0.14	0.12	0.08	0.06	0.05	0.03	0.01
Weight % > than	10	15	20	35	60	95	100
Velocity, in mm/sec	2.11	1.55	0.69	0.39	0.27	0.10	0.01

Plot those data on a chart of weight percent on the y-axis versus particle settling velocity on the x-axis, as follows, then overlay the vertical and horizontal lines shown such that the area outside the resulting boxes above the curve equals the area inside the box and below the curve, for each box. The limiting lines are the calculated v_s and its corresponding weight percentage.

To determine the overall settling basin efficiency, recognize that all particles with a settling velocity greater than v_s will be captured. In this case, that weight percentage is $(1 - 0.51)$, or 49 percent, of all the particles. The question is only what percent of the remainder are likely to be captured. That value is determined by integrating the space above the curve and below the 51 percent line. That is most easily done graphically by drawing the horizontal and vertical lines shown to create boxes that reasonably represent the total volume above the curve. The more boxes that are drawn, the more accurate the results will be, but the longer the calculation will take.

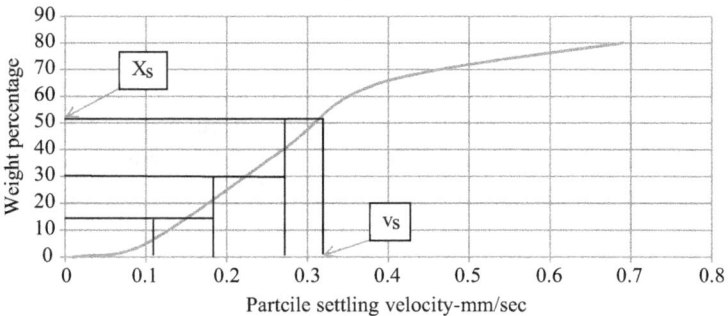

The dimensions of the boxes are the velocity at the end of the box and the difference between the weight percent at the bottom of the box and the weight percent at the top, or the delta weight percentage for that box.

The calculation is done using Equation 4.7 as follows:

$$f = (1 - Xs) + (1/v_s) \Sigma (v\, dx)$$
$$f = (1 - 0.51) + (1/0.32) [(0.12)(0.15\%) + (0.19)(0.15\%) + (0.275)$$
$$(0.21\%)$$

$f = (0.49) + 3.125 (0.018 + 0.029 + 0.058)$

$f = 0.818 = 82\%$ capture of all particles

It is noted that depth never enters these calculations. That means that depth is not as important as the SOR. Thus, if the depth is cut in half, but the flow and other dimensions stay the same, the efficiency of the unit, measured as the total percentage of particles captured, increases because the SOR stays the same, but the design settling velocity will decrease. This is the concept behind plate settlers and tube settlers where the settling distance inside the tubes or between the plates is extremely short, thus allowing for particles with very slow settling velocities to be captured.

4.8   DESIGN OF FLOCCULANT PARTICLE CLARIFIERS

The type of particle discussed in the previous section, the discrete particle, is akin to a sand particle or a crystalline precipitate particle. Chemical treatment, however, yields a flocculant particle that has quite different characteristics from a discrete particle. Therefore, the flocculant particle also settles out differently. This section examines the design of flocculant particle clarifiers.

The main difference between these two types of analyses is that with discrete particles, the SOR is determined, and then the removal efficiency of the clarifier is calculated as a function of that overflow rate. With flocculent particle analysis, the desired removal efficiency is determined, and then the overflow rate required to achieve that efficiency is calculated as a function of the desired efficiency. Unlike discrete particle clarifiers, the removal efficiency of a flocculant particle clarifier is a function of time and depth. The depth, however, is usually used as a trade-off against surface area, since settling time is the key to effective flocculent particle removal.

4.8.1   SETTLING TESTS

The time it takes for specific flocculent particles to settle a certain distance in the water column is determined by conducting a settling test on a sample of the wastewater to be treated. The test is done by putting a well-mixed sample of the wastewater to be treated into a column, at least 300 mm (or about 12 in) diameter and at least as deep as the greatest depth to

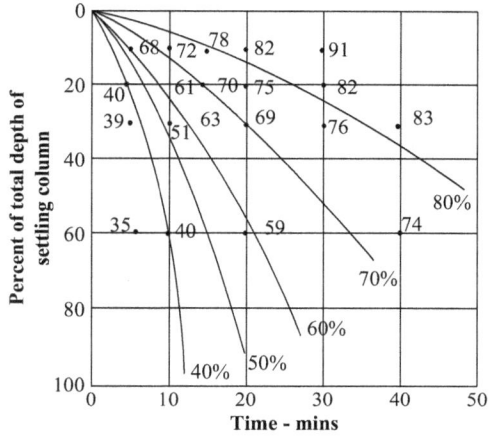

Figure 4.3. Hypothetical settling efficiency plotted as a percentage of particles removed by depth as a function of time.

which the clarifier can be realistically constructed. The rate of settling is then observed over time with samples collected at various depths within the column to determine the suspended solids concentration at that depth and time. Settling efficiency is then plotted as a percentage of particles removed by depth as a function of time. See Figure 4.3.

After the efficiencies have been plotted as a function of time and depth, equipotential lines, or lines of equal removal efficiency are drawn, much like the contours on a contour map of the earth. The design overflow rate is determined by selecting the overall removal efficiency of the reactor for any depth or retention time combination and selecting the one that best suits the site or plant conditions. This is done by drawing a line vertically at the desired detention time and a second line horizontally at the desired depth. The analysis takes the form of summing the average percentage removal between lines, then multiplying that sum by the percentage of particles represented by that removal efficiency interval.

Example Problem 4.3 for flocculant particle settling

Given the following data, determine the total removal efficiency of a flocculant particle sedimentation basin designed for a depth of 7 feet and a detention time of 30 minutes.

Depth Ft	Percent Removal at Stated Time						
	(Time in Minutes)						
	5	10	15	20	25	30	35
0							
2	30	58	75	85	92	95	
4	20	32	50	67	78	88	95
6		30	40	51	65	75	89
8		22	33	45	55	68	80
10		20	30	40	52	62	73

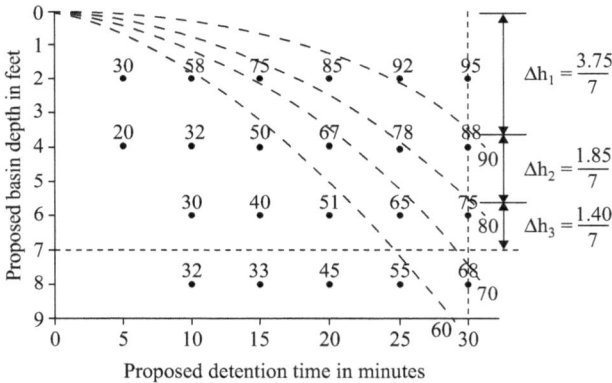

Once the removal efficiencies are plotted, lines of equal removal efficiency are drawn in, as shown in the diagram. A vertical line is then drawn at the desired detention time for the basin under design (30 minutes in this case) and a horizontal line is drawn for the proposed depth of the basin (7 feet in this case). The Δh intervals are then set, starting at the top, or 100 percent removal line, to where the various equal percentage removal lines cross the vertical line at the desired detention time. Here, Δh_1 goes from the 100 percent removal line down to where the 90 percent removal line crosses the 30 minute vertical line at approximately 3.75 feet of depth. That interval is then designated as being equal to 3.75/7. The second interval goes from where the 90 percent equal removal line crosses the 35 minute vertical line to where the 80 percent line crosses at depth of approximately 5.6 feet. The interval depth is the total 5.6 feet minus the first 3.75 feet, or 1.85/7. The third interval goes from where the

70 percent equal removal percentage line crosses the 30 minute vertical line or where the desired depth line intersects the vertical time line. In this case, the depth line at 7 feet crosses the 30 minute line before the 70 percent equal removal efficiency line does. Thus, the distance Δh_3 is calculated as the remaining 1.4/7. The sum of the intervals must equal the total depth of the desired basin.

The total removal efficiency projected for this basin is the sum depth percentages times the average removal efficiency for that interval. Here, 3.75 is the first depth interval and the average removal efficiency is [(100% + 90%)/2] = 95%. For the second interval, the depth is 1.85 and the average removal efficiency is [(90% + 80%)/2] = 85%. For the third interval the depth is 1.4/7 and the average removal efficiency percentage is [80 + 72)/2] = 76%. The average removal efficiency here does not use the 70 percent removal line because the depth line intersects the time line before the 70 percent efficiency line is reached. The intersection point is estimated at 72 percent, in this case, so the average removal efficiency across that interval is the average of 72 percent and 80 percent, or 76 percent.

The calculation of total removal efficiency is then done as follows:

E = (3.75/7) (95%) + (1.85/7) (85%) + (1.4 /7) (76%)
E = 50.9% + 22.5% + 15.2%
E = 88.6%

4.9 HINDERED SETTLING

There is also a third type of settling of concern with respect to wastewater treatment. This is called hindered settling.

Hindered settling occurs when particles falling through the water column begin to get so big that they tend to sweep smaller particles out of the water column ahead of them. That does a nice job of removing extra particles from the water or wastewater, but it tends to slow the rate of settlement of the particles, or "hinders" them from falling freely. Hence the name hindered settling. Hindered settling is most noticeable at the bottom of the settling zone where the large floc tend to build up in precarious piles, much like a proverbial house of cards, until enough weight has been piled up to start to collapse the bridges underneath. Eventually, the bridges collapse and the solids thicken a bit at the bottom.

4.10 DESIGN OF UPFLOW CLARIFIERS

The discussing up until now has revolved around the concept of a rectangular clarifier or sedimentation basin and the concepts of settling that revolve around the particles entering at the end of the clarifier and moving toward the other end, falling out of the water column as they go. There is another kind of clarifier, however, called a circular clarifier, or an upflow clarifier. See Figure 4.4 for a diagram of an upflow clarifier.

With an upflow clarifier, the liquid enters the device inside a shroud at the top of the unit and is swirled around in a countercurrent fashion to slow down the liquid and destroy its energy. As the liquid leaves the bottom of the shroud and starts to travel upward along the outside, it slows down even more and the particles start to settle back down. The settling velocity of the particles still has to exceed the rising velocity of the liquid in order for the particles to settle, and that velocity is still calculated as the influent quantity divided by the surface area of the clarifier, and it is still called the SOR. In this case, however, the area is the round area of the clarifier, inside the effluent baffle, minus the area of the top of the shroud.

Notice in Figure 4.4 that the area of the clarifier continually gets bigger as the flow moves downward inside the shroud and also increases as the flow moves upward again outside the shroud. The larger the flow area becomes, the slower the velocity becomes for the same flow. The clarified flow goes over the effluent weir behind the baffle, while the sludge collects outside the shroud, but inside the baffle, and at the bottom of the

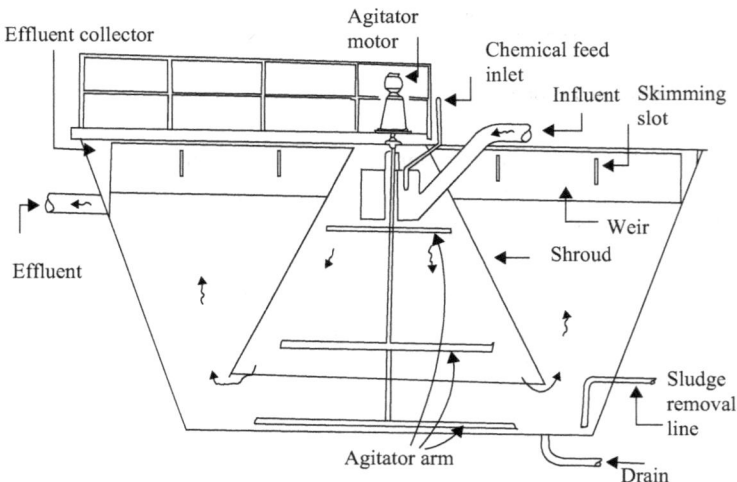

Figure 4.4. Schematic of a typical upflow clarifier.

clarifier. The sludge is periodically removed from the bottom of the clarifier by opening the discharge valve and allowing water pressure inside the clarifier to flush the sludge out.

BIBLIOGRAPHY

Davis, M.L. 2011. *Water and Wastewater Engineering: Design Principles and Practice.* New York, NY: McGraw-Hill Book Co.

Droste, R.E. 1997. *Theory and Practice of Water and Wastewater Treatment.* New York, NY: John Wiley & Sons, Inc.

Hammer, M.J. 2008. *Water and Wastewater Technology.* Upper Saddle River, NJ: Pearson Prentice Hall.

Metcalf & Eddy. 2003. *Wastewater Treatment Engineering: Treatment and Reuse.* 4th ed. New York, NY: McGraw-Hill Publishers.

Metcalf & Eddy/AECOM. 2014. *Wastewater Engineering Treatment and Resource Recovery.* New York, NY: McGraw-Hill Publishers.

CHAPTER 5

SUBSURFACE WASTEWATER DISPOSAL

5.1 INTRODUCTION

There are two places wastewater can go: to a municipal wastewater treatment system through a series of street sewers, or to an on-site, subsurface disposal system. On-site disposal systems, typically in the form of leaching systems, include the soil pipe leaving the building, a septic tank, a connecting pipe to a distribution box, and a series of laterals that distribute the waste into the subsurface soils. In many systems, the laterals are replaced by various forms of chambers and other unique dispersal systems. Newer greywater disposal systems also use small diameter plastic distribution pipes to drip residual water into subsurface environments. Local regulations vary widely with respect to subsurface wastewater disposal and should be carefully reviewed prior to the design and installation of any such system.

5.2 CONVENTIONAL SUBSURFACE DISPOSAL SYSTEMS

Figure 5.1 shows the general arrangement of a septic tank and leaching field. The soil pipe that removes the waste from the building is designed in a standard fashion. The septic tank is normally a concrete or fiberglass prefabricated tank of standard size for various flow rates. They are fundamentally settling basins designed to allow heavy solids and floating matter to separate from the liquid portions of the waste so that a relatively clear liquid can be put into the soil for final cleansing. The distribution box is also typically concrete with sufficient outlets to allow one outlet per distribution pipe. The intent of this unit is to ensure equal flow to all distribution

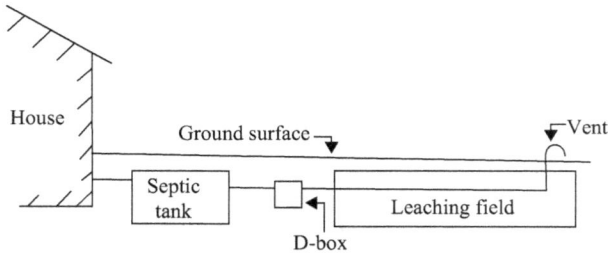

Figure 5.1. Schematic of a typical septic tank and leaching field arrangement.

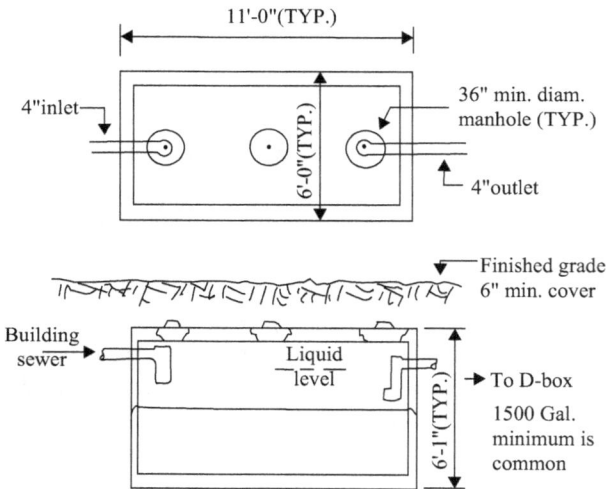

Figure 5.2. Standard septic tank (not to scale).

pipes. These are followed by 4 inch diameter (10 cm) perforated plastic or concrete distribution pipes, set with the perforations on the lower portion of the pipe.

Figure 5.2 illustrates the design of a standard septic tank. The inlet is slightly higher than the outlet. That is very important to prevent back-up of sewage into the building. That means that the tank has a front and a back and that it has to be installed correctly and level, or it will not work properly. There is at least one entry hatch at the top of the tank. It is necessary to clean a septic tank regularly and to provide access for repairing it if needed. Most tanks actually have two hatches, one over the inlet baffle and one over the outlet baffle.

Figure 5.3 describes the general details of a distribution box (D-Box) in a typical distribution system. The inlet is slightly above the outlet to ensure smooth flow and the number of knock-outs for connections varies with the size of the D-Box. Figure 5.4 shows a typical leaching trench detail. Note the variability in the depth of the various backfill components. Many local plumbing codes specify specific depths of each layer as a function of average winter temperatures in their area.

Septic tanks are designed to provide about 48 hours of retention for the expected average daily flows. There are lots of different tables available to help determine average daily flow rates to be expected. In fact, most municipal and state plumbing codes have their own table. In the Commonwealth of Massachusetts that table is found in Title V of the State Environmental Code, for example. Table 5.1 shows typical values similar

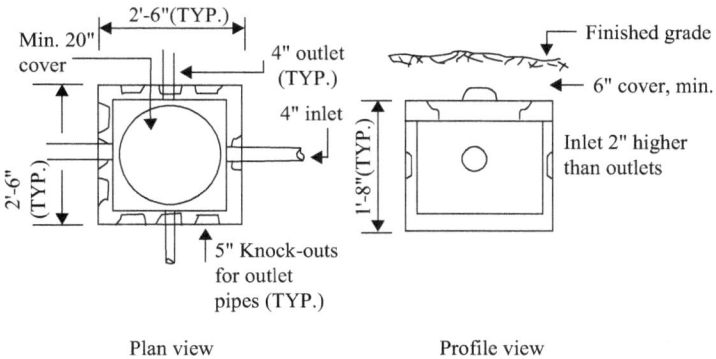

Plan view   Profile view

Figure 5.3. Distribution box (D-Box) (typ.) (not to scale).

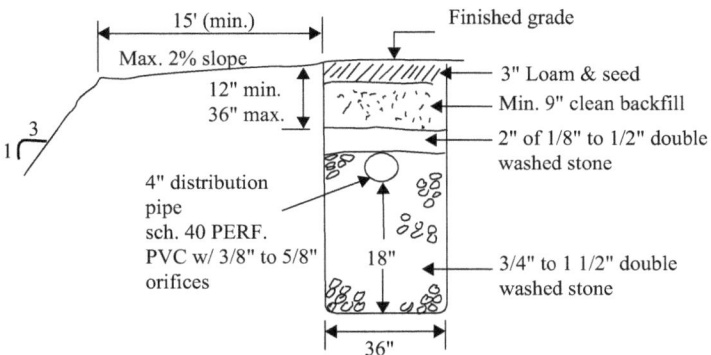

Figure 5.4. Typical leaching trench (not to scale).

Table 5.1. Wastewater design flows from various sources for subsurface disposal systems

Type of establishment	Unit of measure	Gallons per day— gpd	Minimum allowable gpd for system	Liters per day
Residential				
Bed and Breakfast	Per bedroom	110	440	29.1
B & B restaurant open to public—add	Per seat	35	1,000	9.3
Camp, resident, with mess hall, washroom, and toilets	Per person	35		9.3
Camp, day, with washroom and toilets	Per person	10		2.6
Camp, day, with mess hall, washroom, and toilets	Per person	15		4
Campground, with showers and toilets	Per site	90		23.8
Single family dwelling, condo, or cooperative	Per bedroom	110	330	29.1
Multiple family dwelling	Per bedroom	110		29.1
Single family and multiple family dwellings, condos, or apartments	Per person	60	330	226.8
Family mobile home park	Per mobile home	300		79.4
Retirement mobile home park	Per site	150		39.7
Housing for the elderly	Per one or two bedroom unit	150		39.7
Housing for the elderly with more than two bedrooms	Per bedroom	110		29.1

(Continued)

Table 5.1. (*Continued*)

Type of establishment	Unit of measure	Gallons per day— gpd	Minimum allowable gpd for system	Liters per day
Motel, hotel, or boarding house	Per bedroom	110		29.1
Rooming house	Per person	110		29.1
Work or construction camp	Per person	50		13.2
Commercial				
Airport	Per passenger per day	5	150	1.3
Barber shop or beauty salon	Per chair	100		26.5
Bowling alley	Per alley	100		26.5
Country club dining room	Per seat	10		2.6
Country club snack bar or lunch room	Per seat	10		2.6
Country club lockers and showers	Per locker	20		5.3
Doctor's office	Per doctor	250		66.1
Dentist's office	Per dentist	200		52.9
Factory	Per person	15		4
Factory with cafeteria	Per person	20		5.3
Warehouse or dry storage	Per person	15		4
Warehouse or dry storage with cafeteria	Per person	20		5.3
Industrial plant	Per person	15		4
Industrial plant with cafeteria	Per person	20		5.3
Gasoline station	Per island	75	300	19.8
Gasoline station with service bays, add	Per bay	125		33.1

(*Continued*)

Table 5.1. (*Continued*)

Type of establishment	Unit of measure	Gallons per day— gpd	Minimum allowable gpd for system	Liters per day
Service station (no gas)	Per bay	150	450	39.7
Kennel/veterinary office	Per kennel	50		13.2
Lounge, tavern	Per seat	20		5.3
Marina	Per slip	10	500	2.6
Movie theater	Per seat	5		1.3
Public automatic clothes washer	Per machine	400		105.8
Office building	Per 1,000 sf	75	200	19.8
Retail store (except supermarkets)	Per 1,000 sf	50		13.2
Retail store (except supermarkets)	Per toilet room	1,500		396.8
Restaurant	Per seat	35	1,000	9.3
Throughway service area restaurant	Per seat	150	1,000	39.7
Fast food restaurant	Per seat	20	1,000	5.3
Skating rink	Per seat	5	3,000	1.6
Supermarket	Per 1,000 sf	100		26
Swimming pool	Per person	10		2.6
Tennis club	Per court	250		66.1
Auditorium theater	Per seat	5		1.3
Trailer dump station	Per trailer	75		19.8
Seasonal cottages	Per person	50		13.2
Institutional				
Place of worship	Per seat	3		0.8
Place of worship with kitchen	Per seat	6		1.6
Correctional facility	Per bed	200		52.9
Function hall	Per seat	15		4
Gymnasium	Per participant	25		6.6

(*Continued*)

Table 5.1. (*Continued*)

Type of establishment	Unit of measure	Gallons per day— gpd	Minimum allowable gpd for system	Liters per day
Gymnasium	Per spectator	3		0.8
Hospital	Per bed	200		52.9
Nursing home/rest home	Per bed	150		39.7
Public park toilet	Per person	5		1.3
Public park toilet with bathhouse, showers, and flush toilets	Per person	10		2.6
Daycare facility	Per person	10		2.6
Schools				
Elementary school without cafeteria, gymnasium, or showers	Per person	5		1.3
Elementary school with cafeteria, no gymnasium or showers	Per person	8		2.1
Elementary school with cafeteria, gymnasium, and showers	Per person	10		2.6
Secondary/middle school without cafeteria, gymnasium, or showers	Per person	10		2.6
Secondary/middle school with cafeteria, no gymnasium, or showers	Per person	15		4
Secondary/middle school with cafeteria, gymnasium, and showers	Per person	20		5.3
Boarding school, colleges	Per person	65		17.2

to those found in Title V and other codes. It is important, however, to check local codes for these types of values. If there are no codes available, Table 5.1 may be used to help guide system design.

The leaching field is designed based on the permeability of the soil and the expected average daily flow rates through the system. The permeability of the soil is usually determined with a "percolation test" or "perc test." A percolation test procedure is described later. There are several minor variations of this procedure that are possible and some are required by various local authorities. Verification of local requirements is always prudent before embarking on these sorts of testing procedures. Utilization of the procedure shown will provide a good indication of the probability of success with any other modifications to it, however.

5.2.1 PERCOLATION TEST PROCEDURE

A standard percolation test is conducted by performing the following steps in the sequence shown.

1. Prepare a test hole located within the proposed disposal area which, in the judgment of the investigator, appears to be the most limiting. The test hole should have a diameter of 12 inches, with vertical sides, and be 18 inches deep not including any allowable liners or filter layers on either the bottom or sides.

2. Establish a fixed point at the top or bottom of the test hole from which all measurements will be made.

3. Scratch the bottom and sides of the test hole to remove any smeared soil surfaces, taking care not to significantly change the hole dimensions. Add 2 inches of coarse sand to protect the bottom from scouring, or insert a board or other impervious object in the hole so that water may be poured down or on it during the filling operation. A mesh or perforated liner designed to maintain the test hole dimensions in extremely loose soils while allowing essentially unrestricted flow of water may also be used.

4. Carefully fill the hole with clear water to a minimum depth of 12 inches from the bottom. Maintain this minimum 12 inches or greater water level by adding water as necessary in order to saturate surrounding soils for a period of no less than 15 minutes after first filling the hole.

5. After saturation, let the water level drop to a depth of 9 inches above the bottom of the hole and then measure the length of time in minutes for it to drop from a depth of 9 inches to a depth of

6 inches. If the rate is erratic, the hole should be refilled and soaked until the drop per increment of time is steady. The time for the level to drop from a depth of 9 inches to a depth of 6 inches, divided by three, is the percolation rate in minutes per inch.

6. In certain soils, particularly coarse sands, the soil may be so pervious as to make a percolation test difficult, impractical, and meaningless. In this case, the percolation test may generally be discontinued and a rate of 2 minutes per inch or less can be assumed provided that at least 24 gallons of water has been added to the percolation hole within 15 minutes and it is impossible to obtain a liquid depth of 9 inches.

A variation of this procedure provides for the following modifications after the hole is excavated in accordance with the preceding Step 1.

A 2 inch layer of fine gravel or coarse sand is placed in the bottom of the hole and smoothed out. The hole is then filled with clean water to a depth of 12 inches above the gravel or sand. That depth is maintained for a period of at least 4 hours, and preferably overnight, by continually refilling it. If the hole holds water overnight, the depth is adjusted to 6 inches above the sand, then the depth the water level drops in 30 minutes is recorded. If the hole is empty in the morning, it is refilled to 6 inches and the change in depth is recorded at 30-minute intervals for 4 hours. The drop during the last 30 minute interval is recorded as the percolation rate in terms of minutes per 1 inch of drop.

Regardless of the percolation rate method used, that rate is then used to go into another table, such as Table 5.2, to determine the area of leaching trench needed for the expected average daily flow and the soil type found at the site. If the percolation rate is less than 5 minutes per inch, a septic tank and leaching field will generally prove to be a functional disposal system. At percolation rates above that the type of soil will generally dictate the loading rates. Silty and clayey soils will require a much lower loading rate than sandier soils, as indicated in Table 5.2.

The disposal field is constructed of a 4 inch perforated PVC pipe at a slope of 0.005 ft/ft with individual lines generally less than 100 ft long. The trenches are excavated to a depth of at least 18 inches or to the top of the layer in which the percolation test was run. The bottom of the trench must be at least 5 feet above the seasonal high groundwater table. The trench is generally constructed from 3 feet to 4 feet wide and backfilled to a depth of 12 to 14 inches with gravel before the pipe is placed. The pipe is then backfilled with gravel to 2 inches above the pipe and site soils are used to backfill the rest of the way with a layer of geotextile often placed

Table 5.2. Recommended soil loading rates in gpd/sf (L/m²/d) for various soil types

Percolation rate min/inch	Soil type			
	Sand, loamy sand	Loams, sandy loam	Silt loam, sandy clay loams with less than 27 percent clay, silt	Clays, silty clay loam, sandy clay loam with 27 percent or more clay, clay loams, and silty clays
5 or less	0.74 (30.1)	0.60 (24.4)		
6	0.70 (28.5)	0.60 (2.4.4)		
7	0.68 (27.7)	0.60 (24.4)		
8	0.66 (26.9)	0.60 (24.4)		
10		0.60 (24.4)		
15		0.56 (22.8)	0.37 (15.1)	
20		0.53 (21.6)	0.34 (13.8)	
25		0.40 (16.3)	0.33 (13.4)	
30		0.33 (13.4)	0.29 (11.8)	
40			0.25 (10.2)	
50			0.20 (8.1)	0.20 (8.1)
60			0.15 (6.1)	0.15 (6.1)

1 gal/sf/d = 4.05 cm/d
1 gal/sf/d = 40.69 L/m²/d

on top of the gravel first to minimize the seepage of soil into the gravel layer.

Lateral trenches are placed 6 feet or more on center. The total length of pipe required depends on the trench width, since the product of the trench width times the pipe length must equal the area requirement determined from the table. The sides of the trench are not counted when calculating the area or length of laterals required.

The following example illustrates the use of these concepts for a simple septic field design. Reference to Table 5.1 and Table 5.2 is made where needed.

Example Problem 5.1

Determine the required minimum size of a septic tank and percolation (leaching) field for a new subdivision with 200 projected residents. The average percolation rate in the area of the proposed percolation field was found to be 3.5 min/inch and the soil was classified as sandy loam.

Solution

Using 60 gpcd from Table 5.1, the flow is equal to 60 gpcd × 200 people = 12,000 gal/day. In metric units this is 226.8 L/person/day × 200 people = 45,360 L/day

With a percolation rate of 3.5 min/in, Table 5.2 indicates a recommended loading rate of 0.60 gal/sf/day (or 24.4 L/m²/day) for sandy loam soils.

(12,000 gal/day)/(0.6 gal/sf/day) = 20,000 sf of leaching trench required
(45,360 L/day)/(24.4 L/m²/day) = 1,859 m² of leaching trench required

If the trenches are designed with a width of 3 feet, a minimum length of 20,000 sf/3 ft = 6,667 linear feet of trench are required
1,859 m²/1 m of width = 1,859 meters of trench are required

If the trenches are limited to 100 feet or 30 meters, there would need to be 67 lines using feet measurements and 62 using the metric measurements. That is because the width in feet used is slightly smaller than 1 meter (0.91 m).

If the lines are placed 6 feet on center, with a 3 foot buffer along the outside, the field would be approximately 400 feet wide by 106 feet long, or, using a 2 meter center to center placement, and a 2 meter buffer, the field would be about 125 meters long by 34 meters wide.

The septic tank should be designed to hold 48 hours of flow, or 24,000 gallons (3,200 cf) or 90,720 L (90.7 m³). More than one tank and one field may be used and are often required to be used above certain flow volumes, to increase operational efficiencies, but the total tank volume and field sizes would need to be as calculated previously.

5.3 ALTERNATIVE DISPOSAL FIELD DESIGNS

There are several variations on the disposal field that rely on mounds built up over poor soils, nondischarge systems that rely on evapotranspiration, and others.

5.3.1 MOUND SYSTEMS

A mound system is an engineered subsurface disposal field used when conventional septic system disposal fields are likely to fail due to extremely permeable or extremely impermeable soils, areas with shallow soil cover over porous bedrock, and areas where the depth to groundwater is too shallow for normal disposal field construction.

A minimum depth to the seasonal high groundwater elevation and proper soil permeability are required for proper field performance. If the soil above the groundwater table is otherwise suitable, an additional depth of material of equivalent hydraulic permeability is placed over the underlying natural soil until a depth of at least 5 feet from the bottom of the lateral trench to the top of the highest groundwater elevation is achieved throughout the entire field area. The size of the field area is determined based on the hydraulic conductivity (determined from a percolation test) of the underlying pervious soil. The fill material used for the mound must provide equivalent permeability to that of the underlying soils. Some jurisdictions require more precise characteristics of the fill material to ensure exclusion of organic matter, exclusion of material greater than 2 inches in diameter, and other factors that could inhibit flow and cleansing properties of the soil. Where the mound is used to provide separation from the laterals to an underlying pervious bedrock surface, the cleansing and filtering capabilities of the fill material are of critical importance.

In general, the side slopes of a mounded system must not be steeper than three horizontal to one vertical. A minimum of 15 feet, horizontally, must be provided from the outside edge of the outside trench to the top edge of the slope. Generally, the toe of the slope must also be no closer than 5 feet to any property line.

Mound systems that can be fed by gravity from a septic tank, through a distribution box, which generally means they are located downhill from the septic tank effluent outlet, may generally use a normal distribution box for distribution of the effluent throughout the system. Where the field is located up-gradient from the septic tank outlet, a dosing chamber is used to periodically discharge effluent to the distribution box at the field inlet. Dosing chambers are generally set to operate automatically when flow reaches a critical elevation causing a flushing action to occur in the system with discharge to the dosing chamber.

5.3.2 TIGHT TANKS

When existing subsurface disposal field systems have failed, and no reasonably feasible alternative construction is possible, it is sometimes possible to gain local approval to install a tight tank system. This system requires the installation of a septic tank that has a standard inlet, but no outlet. It is designed to hold wastewater generated at the facility until such time as it can be pumped out by a septage hauler and properly disposed off-site.

A tight tank will generally need to conform to the following requirements.

1. The tight tank should be sized at a minimum of 500 percent of the system sewage daily design flow but in no case less than 2,000 gallons.
2. Audio and visual alarms must be set to activate at 60 percent of tank capacity in a suitably convenient location. Transmission of the alarm signal to a location manned 24 hours per day may be required.
3. The tank should have at least one 36-inch diameter cast iron frame and cover at finished grade constructed so as to eliminate the entrance of surface waters. Permanent suction piping may also be required for cleaning of the tank.
4. The tank should be located so as to provide year-round access for pumping.
5. An operation and maintenance plan will usually need to be implemented, which will include a regular pumping schedule and monitoring of the system to ensure proper operation and maintenance.
6. The tank should be waterproof and watertight and should not be located below the water table without extensive testing to prove the integrity of the tank and design against uplift.
7. Aeration, or some other positive method of odor control, may be required.

5.3.3 GREYWATER SYSTEMS

Greywater is generally defined as being any putrescible wastewater discharged from domestic activities including but not necessarily limited to washing machines, sinks, showers, bath tubs, dishwashers, or other sources except toilets, urinals, and any drains equipped with garbage grinders. The excepted flows are often called "blackwater." Greywater from residential, commercial, and public facilities may often be discharged to separate disposal fields or reused.

5.3.3.1 Soil Absorption System for Greywater

When the total discharge to an on-site subsurface sewage disposal consists entirely of greywater, the minimum soil absorption area for residential

systems, as determined by the results of the percolation test, may generally be reduced by no more than 50 percent. Reductions for commercial
and public facility systems should be determined on a case-by-case basis
depending on the actual flows and reuse options. In a remedial upgrade
of an existing system with no increase in flow, the required separation
between the bottom of the soil absorption system and the high groundwater elevation may often be reduced to a minimum of 2 feet in soils with a
recorded percolation rate of more than 2 minutes per inch or a minimum of
3 feet in soils with a recorded percolation rate of 2 minutes or less per inch.

5.3.3.2 Septic Tanks or Filter for a Greywater System

Greywater systems may include either a septic tank or a filter for the separation of any solids that may find their way into the system. The septic tank
should have a minimum effective liquid capacity of 1,000 gallons for household use. Septic tanks for commercial and public facilities should have a
minimum of at least two thirds the size of a septic tank based on total wastewater design flows. When data do not exist on expected design flows, the
design flow should be based on historical flows (1 year or more) from that
facility, or similar facilities if historical flows are not available from the subject facility, and should be 200 percent of average daily water meter readings or 125 percent of the recorded peak daily flow, whichever is greater.

5.3.4 RECIRCULATING SAND FILTERS

A recirculating sand filter (RSF) is defined as a biological and physical
treatment unit consisting of a bed of sand to which septic tank effluent is
distributed and then collected in a recirculating tank prior to recirculating a portion through the sand bed filter and discharging a portion of the
filtrate to a soil absorption system. These systems are generally required
when a disposal field is designed to serve a facility or facilities with a
design flow of 2,000 gallons per day, or more, and they are located in a
Nitrogen-Sensitive Area, defined as interim or defined wellhead protection areas and nitrogen sensitive embayments or other areas that are designated as nitrogen sensitive by local regulatory authorities.

An RSF must generally meet the following requirements:

a. Effluent discharge concentrations should meet or exceed secondary
   treatment standards of 30 mg/L BOD$_5$ and 30 mg/L of total suspended
   solids (TSS). The effluent pH range should be from 6.0 to 9.0.

b. Total nitrogen concentration in the effluent should not exceed 25 mg/L.

c. System owners should monitor effluent quality quarterly for systems serving a facility with a design flow of less than 2,000 gallons per day and monitor both influent and effluent quality quarterly for systems serving a facility with a design flow of 2,000 gallons per day or greater. Monitoring should be conducted for BOD_5, TSS, pH, and total nitrogen.

d. Recirculating sand filter (RSF) systems should contain all components of a standard on-site wastewater disposal system and be capable of functioning as a conventional system.

e. A pressure distribution system is generally required for all systems serving a facility with a design flow of 2,000 gpd or greater.

f. For systems serving a facility with a design flow of 2,000 gpd or greater, the separation between the bottom of the disposal field and the seasonal high groundwater elevation should be calculated after adding the effect of groundwater mounding to the high groundwater elevation.

5.3.5 NON-DISCHARGE OR EVAPOTRANSPIRATION SYSTEMS

An evapotranspiration system utilizes indigenous plants to take up the discharged wastewater and utilize it for growth. Both the water content and the nutrient content are suitable for plant growth. In addition, direct evaporation of water from the surface of the field is also a significant factor in the success of these systems.

Evapotranspiration systems are often suitable for sites with very shallow soil cover, a shallow depth to groundwater, or a shallow depth to fractured bedrock. Typical locations include very small lots on the top of rock cliffs overlooking bays and harbors along coastal areas where there is insufficient room to locate a traditional subsurface disposal system in suitable soil and groundwater conditions, and the lot is so configured or small as to preclude the use of a mounded system. In general, these systems are considered alternative disposal systems under most local regulations and require special permits for construction and use.

Where allowed, these systems should be designed along the following lines.

The system will utilize a standard septic tank system for settling of solids from domestic waste. The septic tank should be oversized by a factor of 1.5 to 2 and be capable of holding a minimum of 3,000 gallons of wastewater. Since the location for these systems is often on rocky ledges, shallow profile septic tanks are suitable, but subject to local regulation. Rock excavation for installation of a septic tank is possible, but usually very expensive and subject to flotation risks if the rock is not well fractured around the installation.

The disposal field is designed in accordance with standard designs except that the percolation rate is assumed to be zero. Should seepage occur it is not generally a bad thing, unless that seepage is likely to intrude directly into fractured bedrock and infiltrate to local groundwater without soil treatment or discharge over a rock cliff as a polluted seep that will cause nuisance algae growth, noxious odors, or health risks to people or the environment. In those cases, plastic linings in the trench must be provided to prevent seepage into the subsurface below the root zone of the local vegetation. Enough trees, shrubs, and grasses to absorb and utilize the total wastewater flow, including seasonal rainfall, must be provided within sufficient proximity of the disposal trenches to allow uptake of the effluent water. The length and location of trenches, then, must be so designed that landscaping of the lot is advantageously affected.

These systems work well in seasonally restricted locations or where trees, shrubs, and other vegetation grow year round. In colder climates, where winter weather stops vegetation growth, these systems will not function properly during the winter months.

The selection of trees, shrubs, and other vegetation with which to landscape a lot utilizing an evapotranspiration system should be done with care. Local vegetation will generally perform better, but care needs to be taken to ensure that sufficient uptake of disposed water will occur to prevent ponding or soft areas on adjacent lawn areas or adjacent lots.

5.3.6   SMALL-DIAMETER SEEPAGE SYSTEMS

In areas where greywater is separated from blackwater, often through the use of composting toilets and waterless toilet facilities for solid sanitary waste streams, coupled with separate collection systems for greywater, a simpler disposal field arrangement can be employed. In these cases, a greywater management system similar to that described previously is employed to separate any stray solids from the water flow. A standard septic tank is used for this purpose, followed by a filter system of some kind. Where the system can operate entirely by gravity, a sand filter can be used. Otherwise a fabric filter is usually more economical and easier to maintain.

Where a pressure filter is needed, the liquid from the septic tank is pumped periodically through the filter system to ensure the removal of fine particulates that could clog the distribution lines. The filtered liquid is then pumped to a distribution box, which may be somewhat remote from the filter system. The distribution box allows the water to flow through a series of small diameter (1 inch [2.5 cm] or less), perforated, distribution pipes.

The pipes are placed at a shallow depth below the ground surface in narrow (3 inch [7.5 cm] maximum width) trenches. The trenches are not necessarily straight, but follow the contours of the existing ground along lines of equal elevation. The pipes will slope at 0.005 ft/ft for maximum efficiency. Seepage of the effluent through the lines occurs at a rate sufficient to provide constant watering of the adjacent vegetation, while minimizing the opportunity for ponding or soft area development on the surface. Seepage into the ground is generally allowed in accordance with greywater disposal field requirements, so this does not constitute a true nondischarge or evapotranspiration system.

Pipe diameters and lengths are determined based on the expected seepage rate per linear foot of pipe and the anticipated water uptake rate of the adjacent vegetation. A factor of 1.5 to 2.0 times the calculated pipe length is recommended to allow for clogging of some pore openings over time.

This type of system has been used successfully for year-round installations in northern climates with limited winter use, provided that the distribution piping is protected from freezing by burial depth or winter compost cover. It is more often considered a seasonal use system in climates where the winter growth of adjacent vegetation is not expected. If there is sufficient separation between the bottom of the trenches and the seasonal high groundwater, however, winter use is generally acceptable because the system does not rely entirely on vegetation growth for disposal success.

BIBLIOGRAPHY

Bouma, J.J. 1975. "A Mound System for Onsite Disposal of Septic Tank Effluent in Slowly Permeable Soils with Seasonally Perched Water Tables." *Journal Environmental Quality* 4, no. 3, pp. 382–388. doi: http://dx.doi.org/10.2134/jeq1975.00472425000400030022x.

Converse, J.C. 1990. *Small Scale Waste Management Project: Wiscomsin At-Grade Soil Absorption System: Siting, Design and Construction Manual.* Madison, WI: University of Wisconsin–Madison.

Hammer, M.J. 2004. *Water and Wastewater Technology.* 5th ed. Upper Saddle River, NJ: Prentice-Hall, Inc.

LaGro, J.A. 1996. "Designing Without Nature: Unsewered Residential Development in Rural Wisconsin." *Landscape and Urban Planning* 35, no. 1, pp. 1–9. doi: http://dx.doi.org/10.1016/0169-2046(96)00314-3.

Massachusetts, Code of July 22, 2013. The State Environmental Code Title 5: Standard Requirements for the Siting, Construction, Inspection, Upgrade and Expansion of On-Site Sewage Treatment and Disposal Systems and for the Transport and Disposal of Septage." *310 CMR 15.0000*. Boston, MA.

McGhee, T.J. 1991. *Water Supply and Sewerage*. 6th ed. New York, NY: McGraw-Hill Book Company.

Metcalf & Eddy, Inc. 2003. *Wastewater Engineering: Treatment and Reuse*. 4th ed. New York, NY: McGraw-Hill Publishers.

Scutella, R.M. 1991. *How to Plan, Contract and Build Your Own Home*. Blue Ridge Summit, PA: Tab Books.

CHAPTER 6

WATER REUSE

6.1   INTRODUCTION

Water reuse takes many forms and has many definitions that are not always consistent. For example, wastewater that has been treated to secondary standards and disinfected is used as irrigation water, while wastewater that has been treated to high level tertiary standards is used to supplement or supplant drinking water. In between there are various uses of greywater and partially treated wastewater for flushing toilets, filling fire hydrants, and washing vehicles. The sixth edition of Hammer and Hammer provides an extensive chart of the types of recycled water use in practice, coupled with the water quality standards applicable to those uses compiled from various state regulations.

In general, wastewater that is to be reused for any unrestricted purpose, or used for a direct human contact purpose, is going to have to be treated to drinking water standards, or nearly so. Wastewater that is to be used for an indirect human contact purpose, such as irrigation of food crops, will need to be treated to secondary standards and be disinfected prior to reuse. Industrial reuse depends on the nature of the source water and the intended industrial reuse requirements. In general, secondary treatment standards will need to be met at a minimum and ultrafiltration for suspended solids, following nutrient and chemical removal, is common.

6.2   SPECIFIC REUSE OPTIONS

Specific reuse options include potable water, firefighting supply, agricultural reuse, urban garden irrigation, greywater reuse, industrial reuse, and groundwater recharge. Each of these reuse options is described later.

6.2.1 POTABLE WATER

It is reported by Davis (2011) that in areas of scarce potable water "literally hundreds of communities" are recycling wastewater for nonpotable uses and that "a half dozen cities, including El Paso, Texas and Los Angeles, California" "have been recharging potable aquifers" with reclaimed wastewater since as early as at least 1962. Historically, reuse of reclaimed wastewater as a potable water supplement has been through recharge of a groundwater aquifer. This usually occurred because the aquifer was being mined at an unsustainable rate and either wells were going dry due to a severe lowering of the groundwater table, or salt water intrusion was occurring where groundwater elevations had been lowered. Direct reuse of reclaimed wastewater has not been widely practiced due to public perceptions of the unproved risks associated with this practice. According to Hammer and Hammer (2008), 47 percent of the reclaimed wastewater in California in 2002 was used to recharge groundwater and 14 percent of the reclaimed water in Florida was used for this purpose in 2005.

Direct reuse of treated (reclaimed) wastewater for potable use involves significant changes to the way wastewater is treated. Most wastewater treatment facilities currently operating in the United States are providing treatment adequate to protect downstream discharge locations, but they do not provide adequate treatment for potability—nor are they intended to do so. The cost to upgrade existing facilities will continue to discourage the practice until the cost of potable water significantly exceeds the cost of retrofitting. The cost of retrofitting is usually so high that construction of a new facility designed specifically to achieve drinking water standards in treated wastewater will be less costly than retrofitting an older facility. Part of that cost is associated with the need to provide large storage facilities for the treated water. Potable water is being generated at a relatively constant rate 24/7, but the use of that water is highly variable on a daily and hourly basis.

The Orange County Water District in Fountain Valley, California, reportedly uses a very sophisticated advanced water treatment plant to prepare wastewater for indirect reuse through groundwater recharge by seepage through a recharge basin, or by injection well recharge to create a salt water intrusion barrier. According to Metcalf & Eddy (2014), that plant subjects the secondary effluent from the Orange County Sanitary District wastewater treatment plant to additional filter screens, chloramine addition, submerged microscreening, sulfuric acid addition for process pH control, cartridge filtration, reverse osmosis, hydrogen peroxide addition, ultraviolet (UV) advanced oxidation and disinfection, CO_2 stripping for decarbonation, and lime addition for final pH control. That water is then used for recharge of the groundwater through recharge basins, or through injection well mounding for salt water intrusion protection. Any flows that exceed the reverse osmosis system capability are by-passed to the advanced UV system, then treated with sodium bisulfite and discharged to a surface water. Even after all of that, direct reuse of the reclaimed water as a potable water supply is not practiced in Orange County.

Direct reuse is not yet common in the United States. Direct reuse in areas of the world where potable water is otherwise extremely difficult to find in sufficient quantities to maintain a developing economy, such as the Middle East and Sub-Saharan Africa, or where political realities put a supply at risk of disruption from neighboring countries, direct reuse is more critical and important. It is likely that global warming is going to change the economics and politics of direct wastewater reuse as a potable water source in the next decade.

6.2.2 FIREFIGHTING SUPPLY

Many communities have determined that providing water for firefighting from scarce potable water supplies is not an effective use of a valuable resource. Consequently, construction of separate firefighting water supply lines that are supplied by a reservoir of treated wastewater has become a significant reuse source for otherwise wasted wastewater. Where a high pressure firefighting hydrant system already exists in a community, connecting the pressure pumps to a treated wastewater reservoir is a relatively low cost option from which the potable water cost savings will easily pay for the cost of conversion in a few short years. Where a separated system does not exist, creating one can be an expensive and time-consuming process that requires decades to repay from potable water cost savings. Those projects are generally done piecemeal; as new water mains are installed a secondary main is placed parallel to the potable main. Only after enough lengths of connected new piping are in place to justify the cost is conversion to treated wastewater to feed the separate firefighting system considered feasible. Once connected, the firefighting system can continue to be expanded over time until the entire community is totally separated.

Treatment for firefighting purposes generally involves reduction of suspended solids following secondary treatment and disinfection. A pH control may be needed, depending on the treatment methodology at the treatment plant and nutrient removal may be needed to minimize algae growth in the reservoir for systems that are not constantly flushing the reservoir for either firefighting use or algae control. Using reservoir contents for other purposes when not being used for firefighting is a viable option where treated water supplies exceed the firefighting needs on a regular basis.

6.2.3 AGRICULTURAL REUSE

Agricultural reuse of wastewater generally involves the application of treated wastewater to agricultural lands for irrigation. To the extent there are reusable nutrients included in the wastewater, that can be a plus, but the water content is the key element in this reuse purpose. Generally, suspended solids reduction is practiced to avoid fouling nozzles and causing excessive erosion wear on pumps, pipes, and elbows.

If the crops to be grown are human food crops, or products used in creating human food products, increased pathogen control is required and increased reductions of contaminants deemed harmful to human health must be removed prior to reuse. The agricultural uptake of various contaminants and compounds of concern is not well understood, in many cases, and caution requires the removal of those constituents to below concentrations of concern to human health prior to reuse on crop lands.

Agricultural reuse of treated wastewater is controlled by the individual states, for the most part. In general, applications are controlled such that direct infiltration to a shallow or potential aquifer is prohibited, application that has the potential for overland runoff to surface water bodies is prohibited, and application rates that could result in direct human contact, through mists or sprays or from children playing in runoff pools, is also prohibited. Use of crops eaten raw requires treatment to near potable standards, including disinfection to destroy pathogens, in most states, as well.

6.2.4 URBAN GARDEN AND LAWN IRRIGATION

The general topic of urban garden and lawn irrigation also includes some commercial and municipal uses such as park, cemetery and recreation area watering, golf course maintenance, and commercial greenhouse irrigation, as well as home gardens and lawns. In all of these applications, with the possible exception of greenhouse maintenance where nonfood crops are grown (such as flowers and shrubs), the reused water must be treated to exceptional standards of suspended solids removal for aesthetics, pathogen removal for health concerns, and nutrient removal to minimize odors and vectors, such as flies and mosquitos, and to minimize the need for retreatment after storage in treated water reservoirs prior to reuse.

Reuse for areas where there is limited public access or opportunity for direct exposure, such as park land, cemetery, and highway median watering, which is typically done at night without spraying or misting, the degree of treatment does not need to be excessive. Generally, secondary treatment with adequate disinfection, often with a slight chlorine residual (enough to maintain control of pathogens, but not enough to adversely affect the plants being watered—a delicate balance sometimes) is warranted. For areas such as golf courses, where human contact is expected, drying of the course prior to human access is required unless the water is treated to unrestricted use standards.

Where home garden and lawn use is allowed, the reclaimed water must generally meet unrestricted use standards, which include a high degree of pathogen reduction, suspended solids removal, and adequate reduction or removal of contaminants deemed harmful to humans and domestic pets.

Clearly, the possibility of direct exposure by humans dictates the treatment required prior to reuse of reclaimed wastewater for irrigation purposes. The degree to which treatment is required in those cases is generally controlled by the individual states.

6.2.5 GREYWATER REUSE

Domestic greywater, or gray water, is water from bathroom sinks, showers, tubs, and washing machines. It is not water that has come into contact with feces, either from a toilet, from washing diapers, or from any other source. It is essential that the greywater contain nothing toxic, such as bleach, dye, bath salts, cleanser, shampoo with unpronounceable ingredients, or products containing boron, which is toxic to plants. It does, however, often contain traces of dirt, food, grease, hair, and certain household cleaning products that make it look gray in color—hence the term. In spite of the color, it is still a safe and even often beneficial source of irrigation water in a yard. The nutrients in greywater can be a valuable fertilizer source for plants.

The most common way to use greywater is to pipe it directly outside and use it to water ornamental plants or fruit trees. Greywater can be generally used directly on vegetables, other than root vegetables, as long as it is not allowed to touch edible parts of the plants.

Industrial greywater is water from a manufacturing process that may or may not require additional treatment prior to reuse. Industrial greywater is most often reused directly in the manufacturing process, when possible. For example, water used as a final rinse of manufactured parts may contain a concentration of contaminants that is not suitable for direct discharge either to the subsurface or to a sewer system. That water may be reused in the initial washing of the manufactured parts, however, since the concentration of contaminants is still much lower than the concentration on the unwashed parts. This reuse of final rinse water will concentrate the contaminants in the initial wash and reduce the overall cost of treating the wastewater. Using treated wastewater from the first wash for a second or intermediate rinse can also reduce downstream treatment expense by drastically reducing the volume of water to be treated. It is also noted that many of the constituents of industrial wastewater can be beneficially recovered from the wash water generally and the more concentrated the contaminants are, the easier they are to recover.

Reuse of recovered industrial wastewater outside of the manufacturing facility depends in large part on the nature of the industry and the nature of the contaminants. Treatment to remove harmful constituents is almost always necessary prior to reuse for irrigation, cooling or heating, or discharge to groundwater or surface water.

6.2.6 GROUNDWATER RECHARGE

Reclaimed wastewater may be recharged directly into existing groundwater supplies in many states, but always subject to regulatory control. If the groundwater is being used as a drinking water supply, or may potentially be used as a drinking water supply, recharge to the groundwater is strictly regulated. Where groundwater is not being used for drinking purposes and is unlikely to be used for drinking water in the foreseeable future, less stringent treatment of the reclaimed water is often allowed.

Recharge can take the form of direct injection into an underground aquifer through a pumped injection well or a groundwater intercept chamber. It may also be recharged through an underground disposal system similar to a septic system, using perforated pipes or chambers. More commonly, it ends up being surface discharged through spray nozzles into access-controlled wooded areas or fields where slow runoff characteristics allow the water to seep into the ground before it has a chance to intercept surface water sources. Sometimes seepage pits are provided to contain batch discharges until they have a chance to seep through underlying soil profiles.

Groundwater recharge is also a way to augment natural processes that create barriers to other flows or to restore water to an aquifer that is being withdrawn at an unsustainable rate. For example, an area along a barrier island or long peninsula may contain a significant freshwater aquifer at shallow depths that is floating on top of a salt water aquifer beneath it. Unsustainable withdrawals from the freshwater aquifer can allow salt water to intrude into the void and thereby contaminate peripheral wells. Reinjection of treated water into the ground around the periphery of the freshwater aquifer can slow the saltwater intrusion and, in some cases, act to reverse historic trends if initiated soon enough. Some states require uncontaminated cooling water to be discharged back into the ground, following appropriate cooling of that water, specifically to minimize the overdrawing of the aquifer and to minimize saltwater intrusion.

Land disposal of treated wastewater is often classified into three to five categories. Those would include soil discharges, wetland discharges, and subsurface discharges. Wetland discharges are not generally considered a reuse because they are being treated by the wetlands, prior to discharge to the surface water associated with the wetlands, and not being used to recharge a groundwater supply. Where the water supply is a surface water that supports downstream flows, discharge through an associated wetland to augment the river or stream flow may be considered a beneficial reuse, however.

Soil discharge is often subcategorized into slow rate infiltration, high rate infiltration, and overland flow. Slow rate systems utilize standard irrigation systems that may contain lines of soaker hose, lines of hose with small dripping perforations, channels or gullies that flood, or large spray irrigators that rotate around a crop field spraying the reclaimed water directly onto cultivated areas. From there it seeps into the ground through the soil. High rate systems typically use large basins to contain a high discharge rate and the water flows outwardly and downward through the surrounding soil profile to recharge the groundwater. The soil generally acts to further cleanse the wastewater of contaminants with the degree of cleansing of dissolved constituents dependent upon the nature of the soil profile. Overland flow is a surface application of treated wastewater. The flow is allowed to run off along the slope of a hill or small mountain until it soaks into the thin soil layer. Often terracing is done perpendicular to the general flow lines to slow the descent of the water down the slope and allow for better seepage control into the underlying soils. These systems have been used successfully in northern climates, such as Greenville, Maine, for example, where a large holding lagoon retains the treated wastewater over the winter, due to slow treatment during that period of time. Discharge to the nearby wooded slopes above Moosehead Lake continues year-round, however, with snow and ice accumulating in the woods during the coldest periods. The snow and ice melt in the spring and the process continues and the discharge is not observable in the lake itself, even though it must surely reach there eventually.

BIBLIOGRAPHY

Davis, M.L. 2011. *Water and Wastewater Engineering: Design Principles and Practice.* New York, NY: McGraw-Hill Book Co.

Droste, R.E. 1997. *Theory and Practice of Water and Wastewater Treatment.* New York: John Wiley & Sons, Inc.

Hammer, M.J., and M.J. Hammer, Jr. 2008. *Water and Wastewater Technology.* Upper Saddle River, NJ: Pearson Prentice Hall.

Managers, W.C. 2004. *Recommended Standards for Wastewater Facilities.* Albany, NY: Health Research, Inc., Health Education Services Division.

McGhee, T. 1991. *Water Supply and Sewerage.* 6th ed. New York, NY: McGraw-Hill Book Company.

Metcalf & Eddy. 2003. *Wastewater Treatment Engineering: Treatment and Reuse.* 4th ed. New York, NY: McGraw-Hill Publishers.

Metcalf & Eddy/AECOM. 2014. *Wastewater Engineering Treatment and Resource Recovery.* New York, NY: McGraw-Hill Publishers.

ABOUT THE AUTHOR

Francis Hopcroft has served as professor of civil and environmental engineering at Wentworth Institute of Technology in Boston, Massachusetts since January of 1984. He was principally responsible for the development of an environmental engineering program at Wentworth (now no longer offered) and for the accreditation of that program through the Accrediting Board for Engineering and Technology.

Prior to starting his teaching career, he spent nearly 25 years working in the environmental consulting and regulatory fields. He has been registered as a professional engineer in six states (currently "retired" in three and he declined to renew in two). He remains actively registered in Massachusetts and is a Licensed Site Professional in the Commonwealth of Massachusetts. He worked for the Federal EPA, the former Metropolitan District Commission in Boston, the Northeast Solid Waste Committee, and several environmental consulting firms before joining the faculty at Wentworth. He is active in the Water Environment Federation, the New England Water Environment Association, and the American Society for Engineering Education.

He is the principle author of a text on the management of hazardous waste on construction sites, a contributing author on two additional texts, and a peer reviewer of several others. He holds two U.S. Patents, including one on a unique wastewater treatment system. He has authored numerous technical papers on various civil and environmental engineering subjects that have been presented at technical conferences and have appeared in the proceedings of those conferences.

INDEX

A

Aerobic bacteria, 36
Aerobic lagoons, 116–117
Aerobic ponds, 117
Agricultural reuse, 178
Algae, 37, 39–40
Alternative disposal field designs
  greywater systems, 169–170
  mound systems, 168
  non-discharge/evapotranspiration
    systems, 171–172
  recirculating sand filter (RSF),
    170–171
  small-diameter seepage systems,
    172–173
  tight tanks, 168–169
Anaerobic bacteria, 37
Anaerobic ponds, 117–118
Atoms, reactive characteristics of
  atomic weight, 3–4
  combining weight, 5–6
  equivalent weight, 5–6
  gram atomic weight, 4
  valence, 4–5
Autotrophic bacteria, 36

B

Bacteria
  aerobic, 36
  anaerobic, 37
  autotrophic, 36
  facultative, 37
  heterotrophic, 36

overview, 35–36
Biological decay rate-k, 43–48
Biological growth curve kinetics,
  50–51
Biological nitrification and
  denitrification, 52–53
Biological oxygen demand (BOD),
  42
  formulas, 43
BOD. *See* Biological oxygen
  demand

C

CFR. *See* Code of Federal
  Regulations
Chemical oxygen demand (COD)
  concepts, 26
  relevance, 26–27
Chlorine, disinfection of
  wastewater, 132–133
Coagulation, 24–26
COD. *See* Chemical oxygen
  demand
Code of Federal Regulations
  (CFR), 59
Compounds, 1–3

D

Disinfection of wastewater
  chlorine, use of, 132–133
  ozone, use of, 133–134
  radiation, 134–135
  ultraviolet light, use of, 134

Dissolved air flotation, 123
Dissolved organic carbon (DOC), 27
Dissolved oxygen (DO), 51–52
DO. *See* Dissolved oxygen

E
Elements, 1–3
Emerging chemicals of concern, 31–33
Environmental Protection Agency (EPA), 59, 60, 61
EPA. *See* Environmental Protection Agency

F
Facultative bacteria, 37
Fats, oil, and grease
  concepts, 28
  relevance, 28
Fixed film systems
  biological towers, 101–104
  rotating biological contactors (RBCs), 104–105
  trickling filters, 81–100
Flocculation, 24–26
Fungi, 40

G
Gravity belt thickeners, 123–125
Gravity thickeners, 121–123
Greywater systems
  septic tanks or filter for, 170
  soil absorption system for, 169–170

H
Heterotrophic bacteria, 36

I
Indicator organisms, 41–42
Inorganic chemicals, 8, 9–11
Ions, 8

M
Material balance calculations, 28–31
Microscopic multicellular organisms, 40–41
Milliequivalents, 12–15
Moles, 6–7

N
National Pollutant Discharge Elimination System (NPDES), 59
Nitrogen, 131
Nitrogenous BOD, 48–49
Normality, 6–7
NPDES. *See* National Pollutant Discharge Elimination System

O
Oxidation state, 4–5
Ozone, disinfection of wastewater, 133–134

P
PAOs. *See* Phosphate accumulating organisms
Pathogens, 38–39, 41
Percolation test procedure, 164–167
Phosphate accumulating organisms (PAOs), 131
Phosphorus, 130
Ponds and lagoons
  aerobic lagoons, 116–117
  aerobic ponds, 117
  anaerobic ponds, 117–118
  facultative lagoons, 118–119
  tertiary lagoons, 120
Preliminary wastewater treatment units
  flow equalization, 69–70
  flow measurement, 67–69
  grit removal, 66–67
  raw sewage pumping, 71

screening and shredding, 62, 66
septage receiving stations, 71–73
typical design criteria for, 63–65
Primary sedimentation basins,
73–77
Primary sludge management,
77, 79
Primary wastewater treatment
units
sedimentation basins, 73–77
sludge management, 77, 79
typical design parameters for, 78
Protozoans, 40

R
Radiation, disinfection of
wastewater, 134–135
Radicals, 1–3
Radicals, properties of, 7–8
RBCs. *See* Rotating biological
contactors
Reaction kinetics. *See* Reaction
rates
Reaction rates
first order reactions, 17–18
k, value of, 19–20
second order reactions, 18–19
third and fourth order reactions,
19
zero order reactions, 16
Recirculating sand filter (RSF),
170–171
Rotary drum thickening, 125
Rotating biological contactors
(RBCs), 104–105
RSF. *See* Recirculating sand filter

S
Secondary wastewater treatment
fixed film systems, 80–105
ponds and lagoons, 115–120
suspended growth biological
treatment systems, 105–115
Sedimentation fundamentals

clarifier design, 139, 142
discrete particle, 137–138
discrete particle clarifiers, design
of, 143–150
flocculant particle clarifiers,
design of, 150–153
flocculant particle sedimentation,
138
flocculator clarifiers, 143
hindered settling, 153
typical design characteristics of,
140–141
upflow clarifiers, design of,
154–155
weirs, 138–139
Sludge management. *See also*
Stabilization of sludge
conditioning, 128
dewatering and final disposal,
129
preliminary sludge handling
operations, 120–121
stabilization, 125–128
thickening, 121–125
Soil loading rates, 166
Stabilization of sludge
aerobic digestion, 126–127
anaerobic digestion, 127–128
lime stabilization, 125–126
temperature-phased digestion,
128
Subsurface wastewater disposal
alternative disposal field designs,
167–173
conventional subsurface
disposal systems, 157–159,
164–167
wastewater design flows,
160–163
Suspended growth biological
treatment systems
conventional activated sludge
systems, 105–113
extended aeration, 114–115

pure oxygen systems, 115
tapered aeration and step feed, 113–114

T
Temperature effects, k-rate, 49–50
Tertiary wastewater treatment units
  intermediary sedimentation basins, 129–130
  nutrient removal techniques, 130–131
TOC. *See* Total organic carbon
Total inorganic carbon, 27
Total organic carbon (TOC)
  concepts, 27
  relevance, 27–28

U
Ultraviolet light, disinfection of wastewater, 134
Units of measure, 8, 11–12

V
VFAs. *See* Volatile fatty acids
Viruses, 37
Volatile fatty acids (VFAs), 131

W
Wastewater treatment
  alkalinity, 22, 23–24
  buffering, 22–23
  composition of raw wastewater, 58
  disinfection of, 131–135
  disinfection of wastewater, 131–135
  effluent water quality requirements, 59–62
  ion-combination reactions, 21–22
  nature of wastewater, 56–57
  oxidation-reduction reactions, 20–21
  parameters, 56–62
  pH, 22
  preliminary treatment units, 62–73
  primary treatment units, 73–79
  secondary treatment, 79–120
  sludge management, 120–129
  stages, 55
  tertiary treatment units, 129–131
Water, hardness of, 26
Water reuse
  agricultural reuse, 178
  firefighting supply, 177–178
  greywater reuse, 179–180
  groundwater recharge, 180–182
  potable water, 176–177
  urban garden and lawn irrigation, 178–179

THIS TITLE IS FROM OUR ENVIRONMENTAL ENGINEERING COLLECTION. OTHER COLLECTIONS INCLUDE...

Industrial Engineering

- Industrial, Systems, and Innovation Engineering — William R. Peterson, Collection Editor
- Manufacturing and Processes — Wayne Hung, Collection Editor
- General Engineering — John K. Estell and Kenneth J. Reid, Collection Editors

Electrical Engineering

- Electrical Power — Hemchandra M. Shertukde, Collection Editor
- Communications and Signal Processing — Orlando Baiocchi, Collection Editor
- Computer Engineering Foundations, Currents and Trajectories — Augustus (Gus) Kinzel Uht, Collection Editor
- Electronic Circuits and Semiconductor Devices — Ashok Goel, Collection Editor

Civil Engineering

- Environmental Engineering — Francis Hopcroft, Collection Editor
- Geotechnical Engineering — Hiroshan Hettiarachchi, Collection Editor
- Transportation Engineering — Bryan Katz, Collection Editor
- Sustainable Structural Systems — Mohammad Noori, Collection Editor

Materials Science and Engineering

- Materials Characterization and Analysis — Richard Brundle, Collection Editor
- Materials Properties and Behavior
- Computational Materials Science
- Nanomaterials

Momentum Press is actively seeking collection editors as well as authors. For more information about becoming an MP author or collection editor, please go to http://www.momentumpress.net/contact

---

Announcing Digital Content Crafted by Librarians

Momentum Press offers digital content as authoritative treatments of advanced engineering topics by leaders in their fields. Hosted on ebrary, MP provides practitioners, researchers, faculty, and students in engineering, science, and industry with innovative electronic content in sensors and controls engineering, advanced energy engineering, manufacturing, and materials science.

Momentum Press offers library-friendly terms:

- perpetual access for a one-time fee
- no subscriptions or access fees required
- unlimited concurrent usage permitted
- downloadable PDFs provided
- free MARC records included
- free trials

The **Momentum Press** digital library is very affordable, with no obligation to buy in future years.

For more information, please visit **www.momentumpress.net/library** or to set up a trial in the US, please contact **mpsales@globalepress.com**.

www.ingramcontent.com/pod-product-compliance
Lightning Source LLC
Chambersburg PA
CBHW070714220326
41598CB00024BA/3146